PIPEFIT...

HANDBOOK

FORREST R. LINDSEY

Martino Publishing
Mansfield Centre, CT
2012

Martino Publishing
P.O. Box 373,
Mansfield Centre, CT 06250 USA

www.martinopublishing.com

ISBN 978-1-61427-329-5

© *2012 Martino Publishing*

Cover design by T. Matarazzo

Printed in the United States of America On 100% Acid-Free Paper

PIPEFITTERS HANDBOOK

FORREST R. LINDSEY

THE INDUSTRIAL PRESS, 93 Worth St., New York 13, N. Y.

Publishers of
AIR CONDITIONING, HEATING AND VENTILATING

PREFACE

This handbook is, to a large extent, the working note-book of the author. It is based on material collected and used by him over a period of thirty-five years of practical experience as a pipefitter. The notes, brought up to date and carefully rearranged, include worked-out tables for pipe bends, offsets, mitered joints, pipe and fitting dimensions, and a host of miscellaneous information useful to the fitter in the shop and in the field.

In many cases, the solution of a practical problem is up to the fitter on the job; or the journeyman may want to check dimensions for shop fabrication of a pipe assembly or a template. The data in this book are intended to give quick answers to these everyday problems as well as to help in the unusual situation.

As a result of his own experience, the author suggested that this book be printed in large type easily read. This has been done, and even though the book may become badly smudged with use, it is believed the tables and text will be quite legible.

The author is indebted to Clifford Strock, editor, and William B. Foxhall, associate editor, of *Air Conditioning Heating and Ventilating* for painstaking arrangement and for data they have contributed, as well as for their deep interest in the production of the book.

FORREST R. LINDSEY

CONTENTS

PIPE BENDING .... 1 to 89
   Setback and Length of Bend .... 2 to 53
   Offset and Crossover Bends .... 54 to 78
   Expansion U-Bends 79
   90° Turns 81
   Coils 85
   Minimum Bending Radius 88
   Wrinkle Bends 89

SCREWED OFFSETS .... 90 to 100
   Common Offset Connections .... 90 to 91
   Travel and Run .... 92 to 97
   Rolling Offset 98
   Screwed Turns 99
   Screwed Coils 100

MITERED JOINTS .... 101 to 129
   Welded Turns .... 102 to 113
   Mitered Coils 114
   Mitered Brackets .... 115 to 120
   Templates for Mitering .... 121 to 129

PIPE DIMENSIONS .... 130 to 137
   Iron and Steel Pipe .... 130 to 133
   Copper Tube .... 134 to 137

SCREWED FITTINGS .... 138 to 172
   Threading Pipe 138
   Standard Threads 139
   Fitting Dimensions .... 141 to 157
   Clearance for Turning .... 158 to 162
   Laying Lengths .... 163 to 172

BUTT WELDING FITTINGS .... 173 to 192
   Elbows, Tees and Crosses .... 173 to 186
   Valves .... 187 to 192

(Continued on next page)

CONTENTS

(Continued from preceding page)

FLANGES AND FLANGE FITTINGS . . . . 193 to 269
   Drilling Flanges . . . . 193 to 199
   Fitting Dimensions . . . . 200 to 243
   Valve Dimensions . . . . 244 to 259
   Flanges and Gaskets . . . . 260 to 269

SOLDER JOINT FITTINGS . . . . 270 to 290
   Soldering and Brazing . . . . 287 to 290

U-BOLTS 291

SPACING OF HANGERS 292

WATER IN PIPES 294

PIPE EXPANSION 296

CONTENTS OF TANKS 297

IDENTIFICATION OF PIPING SYSTEMS . 300

VALVES 304

PLASTIC PIPE 309
   Dimensions and Pressure Limits 313
   Characteristics 325

METALS . . . . 334 to 337
   Spark Tests 334
   Sheet Metal Weights 336
   Weights and Melting Points 337

STEAM 338

MATHEMATICAL DATA . . . . 339 to 364
   Circumferences and Areas of Circles . . 339 to 344
   Squares and Square Roots . . . . 345 to 358
   Areas and Volumes . . . . 359 to 363
   Reciprocals of Numbers 364

WEIGHTS AND MEASURES . . . . 365 to 367

(Concluded on next page)

CONTENTS

(Concluded from preceding page)

CONVERSION TABLES .... 368 to 379
   Inches to Decimals of a Foot 368
   Fahrenheit-Centigrade .... 369 to 373
   Feet of Water to psi 374
   Psi to Feet of Water 375
   Cubic Feet to Gallons 376
   Gallons to Cubic Feet 377
   Miscellaneous Conversion Factors 378
   Useful Formulas 380

TRIGONOMETRY .... 381 to 392
   Trigonometric Functions .... 381 to 385
   Right Triangle Formulas .... 386 to 387
   Solving Problems .... 388 to 392

FIELD LAYOUT OF ANGLES .... 393 to 395
   Angles on the Steel Square 393
   Angles on the Six-foot Rule 395

PIPEFITTERS DICTIONARY .... 396 to 403

Length of Pipe in Bend

Length of pipe in any bend depends on:

(1) Length of bending radius;

(2) Number of degrees of bend

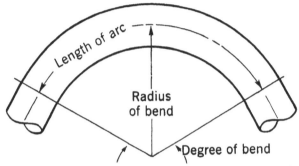

The length of arc in a pipe bend is measured along the centerline of the pipe, and the radius is also measured as extending to the centerline.

In any bend,

Length of Arc = No. of Degrees × .0175 × Radius

Length of Pipe in Standard Bends			
When Angle in Degrees Is	**Multiply Radius by**	**When Angle in Degrees Is**	**Multiply Radius by**
22½	.393	150	2.618
30	.524	180	3.142
45	.785	210	3.665
60	1.047	240	4.189
90	1.571	270	4.712
112½	1.963	300	5.236
120	2.094	360	6.283
135	2.356	540	9.425

Setback and Length of Bend

Setback is a measurement used to locate the beginning of a bend in a piece of pipe. There is a setback measurement for each angle and radius of bend. In a piping layout, setback is the distance from the beginning of the bend to the point where extended centerlines of straight pipe on each leg of the bend would meet.

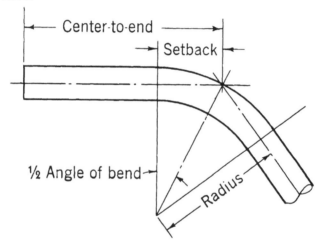

In the sketch above, note that the setback is measured along the center line, as is the length of bend.

Calculated values for setback and length of bend at standard angles, various radii, are given in the tables which follow. Cases not covered by the tables can be calculated by use of trigonometry. As explained in the section dealing with that subject, the formula to use would be:

Setback = Radius × Tangent ½ Angle of Bend

Values of Tangent ½ Angle of Bend for common angles are: 22½°, .199; 30°, .268; 45°, .414; 60°, .577, and 90°, 1.000.

Setback and Length of 15° Bends

Setback and length of bend for 15° bends are calculated from the formulas:

Setback = Radius × .132
Length of Bend = Radius × .262

Calculated values for radius 1 to 72 inches are given in the tables beginning next page.

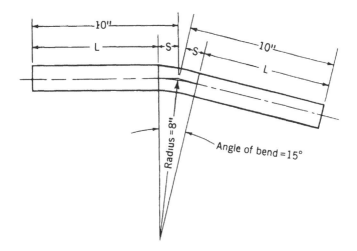

EXAMPLE

Find the length of straight pipe required to make a 15° bend on 8-inch radius so that the distance from each end of the pipe to intersection of centerlines of straight legs is 10 inches.

SOLUTION

(1) In table on the next page, for 8-inch radius: Setback = 1$\frac{1}{16}$ inches; Length of Bend = 2$\frac{1}{16}$ inches

(2) Straight leg (marked L in drawing) is 10 inches less the setback (marked S in drawing); therefore

L = 10 — 1$\frac{1}{16}$ = 8$\frac{15}{16}$ inches

(3) Total length of pipe in piece is

(2× L) + Length of Bend = (2 × 8$\frac{15}{16}$) + 2$\frac{1}{16}$
= 19$\frac{15}{16}$ inches

Setback and Length of 15° Bends, Radius 1 to 72 Inches

Radius of Bend, Inches	Setback, Inches	Length of Bend, Inches
1	$\frac{1}{8}$	$\frac{1}{4}$
2	$\frac{1}{4}$	$\frac{1}{2}$
3	$\frac{3}{8}$	$\frac{13}{16}$
4	$\frac{1}{2}$	$1\frac{1}{16}$
5	$\frac{11}{16}$	$1\frac{5}{16}$
6	$\frac{13}{16}$	$1\frac{9}{16}$
7	$\frac{15}{16}$	$1\frac{13}{16}$
8	$1\frac{1}{16}$	$2\frac{1}{16}$
9	$1\frac{3}{16}$	$2\frac{3}{8}$
10	$1\frac{5}{16}$	$2\frac{5}{8}$
11	$1\frac{7}{16}$	$2\frac{7}{8}$
12	$1\frac{9}{16}$	$3\frac{1}{8}$
13	$1\frac{11}{16}$	$3\frac{7}{16}$
14	$1\frac{7}{8}$	$3\frac{11}{16}$
15	2	$3\frac{15}{16}$
16	$2\frac{1}{8}$	$4\frac{3}{16}$
17	$2\frac{1}{4}$	$4\frac{7}{16}$
18	$2\frac{3}{8}$	$4\frac{11}{16}$
19	$2\frac{1}{2}$	5
20	$2\frac{5}{8}$	$5\frac{1}{4}$
21	$2\frac{3}{4}$	$5\frac{1}{2}$
22	$2\frac{7}{8}$	$5\frac{3}{4}$
23	$3\frac{1}{16}$	6
24	$3\frac{3}{16}$	$6\frac{5}{16}$

Setback and Length of 15° Bends, Radius 1 to 72 Inches

Radius of Bend, Inches	Setback, Inches	Length of Bend, Inches
25	$3\frac{5}{16}$	$6\frac{9}{16}$
26	$3\frac{7}{16}$	$6\frac{13}{16}$
27	$3\frac{9}{16}$	$7\frac{1}{16}$
28	$3\frac{11}{16}$	$7\frac{5}{16}$
29	$3\frac{13}{16}$	$7\frac{5}{8}$
30	$3\frac{15}{16}$	$7\frac{7}{8}$
31	$4\frac{1}{16}$	$8\frac{1}{8}$
32	$4\frac{1}{4}$	$8\frac{3}{8}$
33	$4\frac{3}{8}$	$8\frac{5}{8}$
34	$4\frac{1}{2}$	$8\frac{15}{16}$
35	$4\frac{5}{8}$	$9\frac{3}{16}$
36	$4\frac{3}{4}$	$9\frac{7}{16}$
37	$4\frac{7}{8}$	$9\frac{11}{16}$
38	5	$9\frac{15}{16}$
39	$5\frac{1}{8}$	$10\frac{3}{16}$
40	$5\frac{1}{4}$	$10\frac{1}{2}$
41	$5\frac{7}{16}$	$10\frac{3}{4}$
42	$5\frac{9}{16}$	11
43	$5\frac{11}{16}$	$11\frac{1}{4}$
44	$5\frac{13}{16}$	$11\frac{1}{2}$
45	$5\frac{15}{16}$	$11\frac{13}{16}$
46	$6\frac{1}{16}$	$12\frac{1}{16}$
47	$6\frac{3}{16}$	$12\frac{5}{16}$
48	$6\frac{5}{16}$	$12\frac{9}{16}$

Setback and Length of 15° Bends, Radius 1 to 72 Inches

Radius of Bend, Inches	Setback, Inches	Length of Bend, Inches
49	$6\frac{7}{16}$	$12\frac{13}{16}$
50	$6\frac{5}{8}$	$13\frac{1}{8}$
51	$6\frac{3}{4}$	$13\frac{3}{8}$
52	$6\frac{7}{8}$	$13\frac{5}{8}$
53	7	$13\frac{7}{8}$
54	$7\frac{1}{8}$	$14\frac{1}{8}$
55	$7\frac{1}{4}$	$14\frac{7}{16}$
56	$7\frac{3}{8}$	$14\frac{11}{16}$
57	$7\frac{1}{2}$	$14\frac{15}{16}$
58	$7\frac{11}{16}$	$15\frac{3}{16}$
59	$7\frac{13}{16}$	$15\frac{7}{16}$
60	$7\frac{15}{16}$	$15\frac{3}{4}$
61	$8\frac{1}{16}$	16
62	$8\frac{3}{16}$	$16\frac{1}{4}$
63	$8\frac{5}{16}$	$16\frac{1}{2}$
64	$8\frac{7}{16}$	$16\frac{3}{4}$
65	$8\frac{9}{16}$	17
66	$8\frac{11}{16}$	$17\frac{5}{16}$
67	$8\frac{7}{8}$	$17\frac{9}{16}$
68	9	$17\frac{13}{16}$
69	$9\frac{1}{8}$	$18\frac{1}{16}$
70	$9\frac{1}{4}$	$18\frac{5}{16}$
71	$9\frac{3}{8}$	$18\frac{5}{8}$
72	$9\frac{1}{2}$	$18\frac{7}{8}$

Setback and Length of 22½° Bends

Setback and length of bend for 22½° bends are calculated from the formulas:

$$\text{Setback} = \text{Radius} \times .199$$
$$\text{Length of Bend} = \text{Radius} \times .393$$

Calculated values for radius 1 to 72 inches are given in the tables beginning next page.

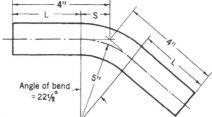

EXAMPLE

It is desired to make a 22½° bend on 5-inch radius so that the distance from each end of the pipe to the point where centerlines of bend legs intersect is 4 inches. Find length of straight pipe in each leg and total length of pipe in piece.

SOLUTION

(1) In table on next page, for 5-inch radius:

Setback = 1 inch; Length of Bend = 2 inches

(2) Length of straight pipe in each leg of bend (marked L in drawing) is:

4 inches — Setback (S) = 3 inches

(3) Length of pipe in whole piece is:

(2 × L) + Length of Bend = (2 × 3) + 2 = 8 inches

(4) On straight piece, measure off 3 inches to mark beginning of bend.

(5) Lay off length of bend, 2 inches, and bend on a 10-inch wheel (5-inch radius) from beginning to end of bend.

(6) If piece is to be flanged, include flange allowance in 3-inch leg.

Setback and Length of 22½° Bends, Radius 1 to 72 Inches

Radius of Bend, Inches	Setback, Inches	Length of Bend, Inches
1	$\frac{3}{16}$	$\frac{3}{8}$
2	$\frac{3}{8}$	$\frac{13}{16}$
3	$\frac{5}{8}$	$1\frac{3}{16}$
4	$\frac{13}{16}$	$1\frac{9}{16}$
5	1	2
6	$1\frac{3}{16}$	$2\frac{3}{8}$
7	$1\frac{3}{8}$	$2\frac{3}{4}$
8	$1\frac{5}{8}$	$3\frac{1}{8}$
9	$1\frac{13}{16}$	$3\frac{1}{2}$
10	2	$3\frac{15}{16}$
11	$2\frac{3}{16}$	$4\frac{5}{16}$
12	$2\frac{3}{8}$	$4\frac{11}{16}$
13	$2\frac{5}{8}$	$5\frac{1}{8}$
14	$2\frac{13}{16}$	$5\frac{1}{2}$
15	3	$5\frac{13}{16}$
16	$3\frac{3}{16}$	$6\frac{1}{4}$
17	$3\frac{3}{8}$	$6\frac{11}{16}$
18	$3\frac{5}{8}$	$7\frac{1}{16}$
19	$3\frac{13}{16}$	$7\frac{7}{16}$
20	4	$7\frac{7}{8}$
21	$4\frac{3}{16}$	$8\frac{1}{4}$
22	$4\frac{3}{8}$	$8\frac{5}{8}$
23	$4\frac{5}{8}$	$9\frac{1}{16}$
24	$4\frac{13}{16}$	$9\frac{7}{16}$

Setback and Length of 22½° Bends, Radius 1 to 72 Inches

Radius of Bend, Inches	Setback, Inches	Length of Bend, Inches
25	5	$9\frac{13}{16}$
26	$5\frac{3}{16}$	$10\frac{3}{16}$
27	$5\frac{3}{8}$	$10\frac{5}{8}$
28	$5\frac{5}{8}$	11
29	$5\frac{13}{16}$	$11\frac{3}{8}$
30	6	$11\frac{7}{8}$
31	$6\frac{3}{16}$	$12\frac{3}{16}$
32	$6\frac{1}{2}$	$12\frac{9}{16}$
33	$6\frac{9}{16}$	$12\frac{15}{16}$
34	$6\frac{3}{4}$	$13\frac{3}{8}$
35	7	$13\frac{3}{4}$
36	$7\frac{3}{16}$	$14\frac{1}{8}$
37	$7\frac{3}{8}$	$14\frac{1}{2}$
38	$7\frac{9}{16}$	$14\frac{15}{16}$
39	$7\frac{3}{4}$	$15\frac{5}{16}$
40	8	$15\frac{3}{4}$
41	$8\frac{3}{16}$	$16\frac{1}{8}$
42	$8\frac{3}{8}$	$16\frac{1}{2}$
43	$8\frac{9}{16}$	$16\frac{7}{8}$
44	$8\frac{3}{4}$	$17\frac{1}{4}$
45	9	$17\frac{5}{8}$
46	$9\frac{1}{8}$	$18\frac{1}{16}$
47	$9\frac{3}{8}$	$18\frac{1}{2}$
48	$9\frac{9}{16}$	$18\frac{7}{8}$

Setback and Length of 22½° Bends, Radius 1 to 72 Inches

Radius of Bend, Inches	Setback, Inches	Length of Bend, Inches
49	9¾	19¼
50	10	19¹¹⁄₁₆
51	10⅛	20
52	10⅜	20⁷⁄₁₆
53	10½	20¹³⁄₁₆
54	10¾	21³⁄₁₆
55	10¹⁵⁄₁₆	21⅝
56	11⅛	22
57	11⁵⁄₁₆	22⅜
58	11⁹⁄₁₆	22¾
59	11¹³⁄₁₆	23⅛
60	11⅞	23⁹⁄₁₆
61	12³⁄₁₆	24
62	12⅜	24⅜
63	12⅝	24¹³⁄₁₆
64	12¹³⁄₁₆	25³⁄₁₆
65	13	25⅝
66	13³⁄₁₆	26
67	13⅜	26⅜
68	13⅝	26¹³⁄₁₆
69	13¹³⁄₁₆	27³⁄₁₆
70	14	27⁹⁄₁₆
71	14³⁄₁₆	27¹⁵⁄₁₆
72	14⅜	28⅜

Setback and Length of 30° Bends

Setback and length of bend for **30°** bends are calculated from the formulas:

$$\text{Setback} = \text{Radius} \times .268$$
$$\text{Length of Bend} = \text{Radius} \times .524$$

Calculated values for radius 1 to **72** inches are given in the tables beginning next **page.**

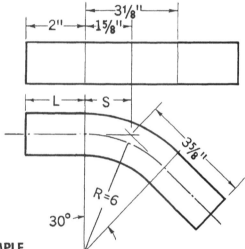

EXAMPLE

Find the length of straight pipe required to make a 30° bend on 6-inch radius so that the distance from each end of the pipe to where centerlines of straight legs meet is $3\frac{5}{8}$ inches.

SOLUTION

(1) In table on next page, for 6-inch radius:
Setback = $1\frac{5}{8}$ inches; Length of Bend = $3\frac{1}{8}$ inches

(2) Straight Leg (marked L in drawing) is $3\frac{5}{8}$ less the setback (S in the drawing), so

$$L = 3\frac{5}{8} - 1\frac{5}{8} = 2 \text{ inches}$$

(3) Total length of pipe in piece is
$2 \times \text{Leg} + \text{Length of Bend} = 4 + 3\frac{1}{8} = 7\frac{1}{8}$ inches

Setback and Length of 30° Bends, Radius 1 to 72 Inches

Radius of Bend, Inches	Setback, Inches	Length of Bend, Inches
1	$\frac{1}{4}$	$\frac{1}{2}$
2	$\frac{9}{16}$	$1\frac{1}{16}$
3	$\frac{13}{16}$	$1\frac{9}{16}$
4	$1\frac{1}{16}$	$2\frac{1}{8}$
5	$1\frac{3}{8}$	$2\frac{5}{8}$
6	$1\frac{5}{8}$	$3\frac{1}{8}$
7	$1\frac{7}{8}$	$3\frac{11}{16}$
8	$2\frac{1}{8}$	$4\frac{1}{8}$
9	$2\frac{7}{16}$	$4\frac{11}{16}$
10	$2\frac{11}{16}$	$5\frac{1}{4}$
11	$2\frac{15}{16}$	$5\frac{3}{4}$
12	$3\frac{1}{4}$	$6\frac{5}{16}$
13	$3\frac{1}{2}$	$6\frac{13}{16}$
14	$3\frac{3}{4}$	$7\frac{3}{8}$
15	4	$7\frac{7}{8}$
16	$4\frac{5}{16}$	$8\frac{3}{8}$
17	$4\frac{9}{16}$	$8\frac{15}{16}$
18	$4\frac{13}{16}$	$9\frac{7}{16}$
19	$5\frac{1}{8}$	10
20	$5\frac{3}{8}$	$10\frac{1}{2}$
21	$5\frac{5}{8}$	11
22	$5\frac{7}{8}$	$11\frac{9}{16}$
23	$6\frac{3}{16}$	$12\frac{1}{16}$
24	$6\frac{7}{16}$	$12\frac{5}{8}$

Setback and Length of 30° Bends, Radius 1 to 72 Inches

Radius of Bend, Inches	Setback, Inches	Length of Bend, Inches
25	$6\frac{11}{16}$	$13\frac{1}{8}$
26	7	$13\frac{5}{8}$
27	$7\frac{1}{4}$	$14\frac{3}{16}$
28	$7\frac{1}{2}$	$14\frac{11}{16}$
29	$7\frac{3}{4}$	$15\frac{3}{16}$
30	$8\frac{1}{16}$	$15\frac{3}{4}$
31	$8\frac{5}{16}$	$16\frac{1}{4}$
32	$8\frac{9}{16}$	$16\frac{13}{16}$
33	$8\frac{7}{8}$	$17\frac{3}{16}$
34	$9\frac{1}{8}$	$17\frac{13}{16}$
35	$9\frac{3}{8}$	$18\frac{3}{8}$
36	$9\frac{11}{16}$	$18\frac{7}{8}$
37	$9\frac{15}{16}$	$19\frac{3}{8}$
38	$10\frac{3}{16}$	20
39	$10\frac{1}{2}$	$20\frac{7}{16}$
40	$10\frac{3}{4}$	21
41	11	$21\frac{1}{2}$
42	$11\frac{1}{4}$	22
43	$11\frac{1}{2}$	$22\frac{1}{2}$
44	$11\frac{13}{16}$	$23\frac{1}{16}$
45	$12\frac{1}{16}$	$23\frac{9}{16}$
46	$12\frac{3}{8}$	$24\frac{1}{8}$
47	$12\frac{5}{8}$	$24\frac{5}{8}$
48	$12\frac{7}{8}$	$25\frac{1}{8}$

Setback and Length of 30° Bends, Radius 1 to 72 Inches

Radius of Bend, Inches	Setback, Inches	Length of Bend, Inches
49	$13\frac{1}{8}$	$25\frac{11}{16}$
50	$13\frac{7}{16}$	$26\frac{3}{16}$
51	$13\frac{11}{16}$	$26\frac{3}{4}$
52	$13\frac{5}{16}$	$27\frac{1}{4}$
53	$14\frac{1}{4}$	$27\frac{3}{4}$
54	$14\frac{1}{2}$	$28\frac{5}{16}$
55	$14\frac{3}{4}$	$28\frac{13}{16}$
56	15	$29\frac{3}{8}$
57	$15\frac{1}{4}$	$29\frac{7}{8}$
58	$15\frac{9}{16}$	$30\frac{3}{8}$
59	$15\frac{3}{4}$	$30\frac{7}{8}$
60	$16\frac{1}{8}$	$31\frac{7}{16}$
61	$16\frac{3}{8}$	32
62	$16\frac{5}{8}$	$32\frac{1}{2}$
63	$16\frac{7}{8}$	$33\frac{1}{16}$
64	$17\frac{3}{16}$	$33\frac{5}{8}$
65	$17\frac{7}{16}$	$34\frac{1}{8}$
66	$17\frac{11}{16}$	$34\frac{11}{16}$
67	18	$35\frac{3}{16}$
68	$18\frac{1}{4}$	$35\frac{9}{16}$
69	$18\frac{1}{2}$	$36\frac{1}{4}$
70	$18\frac{3}{4}$	$36\frac{3}{4}$
71	$19\frac{1}{16}$	$37\frac{1}{4}$
72	$19\frac{5}{16}$	$37\frac{13}{16}$

Setback and Length of 45° Bends

Setback and length of bend for 45° bends are calculated from the formulas:

$$\text{Setback} = \text{Radius} \times .414$$
$$\text{Length of Bend} = \text{Radius} \times .785$$

Calculated values for radius 1 to 72 inches are given in the tables beginning next page.

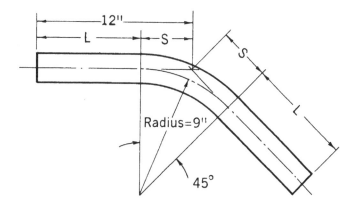

EXAMPLE

Find the distance from pipe end at which to begin a 45° bend on 9-inch radius so that the distance from pipe end to point where centerlines of straight legs intersect is 12 inches. If legs are equal, what length of pipe is used?

SOLUTION

(1) In table on next page, for 9-inch radius:
Setback $= 3\frac{3}{4}$ inches; Length of Bend $= 7\frac{1}{16}$ inches

(2) 12″ − Setback (S in drawing) $= 12 - 3\frac{3}{4} = 8\frac{1}{4}$ inches. Therefore begin bend $8\frac{1}{4}$ inches from end.

(3) If straight legs (L in drawing) are equal,
Length of Piece $= 2L + $ Length of Bend
$$= (2 \times 8\frac{1}{4}) + 7\frac{1}{16} = 23\frac{9}{16} \text{ inches}$$

Setback and Length of 45° Bends, Radius 1 to 72 Inches

Radius of Bend, Inches	Setback, Inches	Length of Bend, Inches
1	$\frac{7}{16}$	$\frac{3}{4}$
2	$\frac{13}{16}$	$1\frac{9}{16}$
3	$1\frac{1}{4}$	$2\frac{3}{8}$
4	$1\frac{5}{8}$	$3\frac{1}{8}$
5	$2\frac{1}{16}$	$3\frac{15}{16}$
6	$2\frac{7}{16}$	$4\frac{11}{16}$
7	$2\frac{7}{8}$	$5\frac{1}{2}$
8	$3\frac{1}{4}$	$6\frac{1}{4}$
9	$3\frac{3}{4}$	$7\frac{1}{16}$
10	$4\frac{1}{8}$	$7\frac{13}{16}$
11	$4\frac{1}{2}$	$8\frac{5}{8}$
12	$4\frac{15}{16}$	$9\frac{7}{16}$
13	$5\frac{5}{16}$	$10\frac{3}{16}$
14	$5\frac{3}{4}$	11
15	$6\frac{1}{8}$	$11\frac{3}{4}$
16	$6\frac{9}{16}$	$12\frac{9}{16}$
17	7	$13\frac{5}{16}$
18	$7\frac{7}{16}$	$14\frac{1}{16}$
19	$7\frac{7}{8}$	$14\frac{15}{16}$
20	$8\frac{3}{16}$	$15\frac{11}{16}$
21	$8\frac{5}{8}$	$16\frac{1}{2}$
22	$9\frac{1}{8}$	$17\frac{1}{4}$
23	$9\frac{7}{16}$	$18\frac{1}{16}$
24	$9\frac{15}{16}$	$18\frac{13}{16}$

Setback and Length of 45° Bends, Radius 1 to 72 Inches

Radius of Bend, Inches	Setback, Inches	Length of Bend, Inches
25	10¼	19 9/16
26	10 11/16	20⅜
27	11 1/16	21 3/16
28	11½	22
29	11⅞	22¾
30	12 7/16	23 9/16
31	12 13/16	24⅜
32	13¼	25⅛
33	13 11/16	25 15/16
34	14 1/16	26¾
35	14½	27½
36	14 15/16	28¼
37	15 5/16	29 1/16
38	15¾	29⅞
39	16⅛	30⅝
40	16 9/16	31 7/16
41	16⅞	32 3/16
42	17⅜	33
43	17 13/16	33¾
44	18¼	34 9/16
45	18⅝	35⅜
46	19 1/16	36⅛
47	19½	36 15/16
48	19⅞	37¾

Setback and Length of 45° Bends, Radius 1 to 72 Inches

Radius of Bend, Inches	Setback, Inches	Length of Bend, Inches
49	20$\frac{5}{16}$	38$\frac{1}{2}$
50	20$\frac{3}{4}$	39$\frac{1}{4}$
51	21$\frac{1}{8}$	40$\frac{1}{16}$
52	21$\frac{9}{16}$	40$\frac{7}{8}$
53	22	41$\frac{5}{8}$
54	22$\frac{3}{8}$	42$\frac{7}{16}$
55	22$\frac{3}{4}$	43$\frac{3}{16}$
56	23$\frac{3}{16}$	44$\frac{1}{16}$
57	23$\frac{5}{8}$	44$\frac{3}{4}$
58	24	45$\frac{9}{16}$
59	24$\frac{7}{16}$	46$\frac{5}{16}$
60	24$\frac{7}{8}$	47$\frac{1}{4}$
61	25$\frac{1}{4}$	48
62	25$\frac{11}{16}$	48$\frac{13}{16}$
63	26$\frac{1}{8}$	49$\frac{5}{8}$
64	26$\frac{9}{16}$	50$\frac{3}{8}$
65	26$\frac{15}{16}$	51$\frac{3}{16}$
66	27$\frac{3}{8}$	52
67	27$\frac{15}{16}$	52$\frac{3}{4}$
68	28$\frac{3}{16}$	53$\frac{1}{2}$
69	28$\frac{5}{8}$	54$\frac{3}{8}$
70	29	55$\frac{1}{16}$
71	29$\frac{7}{16}$	55$\frac{15}{16}$
72	29$\frac{13}{16}$	56$\frac{11}{16}$

Setback and Length of 60° Bends

Setback and length of bend for 60° bends are calculated from the formulas:

$$\text{Setback} = \text{Radius} \times .577$$
$$\text{Length of Bend} = \text{Radius} \times 1.05$$

Calculated values for radius 1 to 72 inches are given in the tables beginning next page.

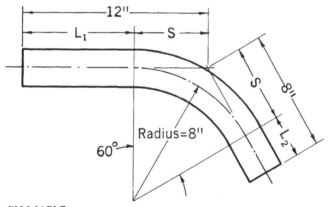

EXAMPLE

Find the length of straight pipe required for a piece including a 60° bend on 8-inch radius such that the distance from one end of the pipe to the point where centerlines of straight legs intersect is 12 inches and from the other end to the same point is 8 inches.

SOLUTION

(1) In table on next page, for 8-inch radius, find Setback = $4\frac{5}{8}$ inches; Length of Bend = $8\frac{1}{4}$ inches

(2) One leg of straight pipe (L_1 in sketch) = 12 − $4\frac{5}{8}$ = $7\frac{3}{8}$ inches. The other leg (L_2) = 8 − $4\frac{5}{8}$ = $3\frac{3}{8}$ inches

(3) Length of pipe required is

$$L_1 + L_2 + \text{Length of Bend} = 7\frac{3}{8} + 3\frac{3}{8} + 8\frac{1}{4}$$
$$= 19 \text{ inches}$$

Setback and Length of 60° Bends, Radius 1 to 72 Inches

Radius of Bend, Inches	Setback, Inches	Length of Bend, Inches
1	$\frac{9}{16}$	$1\frac{1}{16}$
2	$1\frac{1}{8}$	$2\frac{1}{8}$
3	$1\frac{3}{4}$	$3\frac{1}{8}$
4	$2\frac{5}{16}$	$4\frac{3}{16}$
5	$2\frac{7}{8}$	$5\frac{1}{4}$
6	$3\frac{1}{2}$	$6\frac{5}{16}$
7	$4\frac{1}{16}$	$7\frac{5}{16}$
8	$4\frac{5}{8}$	$8\frac{1}{4}$
9	$5\frac{3}{16}$	$9\frac{7}{16}$
10	$5\frac{13}{16}$	$10\frac{7}{16}$
11	$6\frac{3}{8}$	$11\frac{1}{2}$
12	$6\frac{15}{16}$	$12\frac{9}{16}$
13	$7\frac{1}{2}$	$13\frac{5}{8}$
14	$8\frac{1}{8}$	$14\frac{11}{16}$
15	$8\frac{11}{16}$	$15\frac{11}{16}$
16	$9\frac{1}{4}$	$16\frac{3}{4}$
17	$9\frac{7}{8}$	$17\frac{13}{16}$
18	$10\frac{7}{16}$	$18\frac{13}{16}$
19	11	$19\frac{15}{16}$
20	$11\frac{9}{16}$	$20\frac{15}{16}$
21	$12\frac{3}{16}$	22
22	$12\frac{3}{4}$	$23\frac{1}{16}$
23	$13\frac{5}{16}$	$24\frac{1}{16}$
24	$13\frac{15}{16}$	$25\frac{1}{8}$

Setback and Length of 60° Bends, Radius 1 to 72 Inches

Radius of Bend, Inches	Setback, Inches	Length of Bend, Inches
25	14½	26³⁄₁₆
26	15¹⁄₁₆	27³⁄₁₆
27	15⅝	28⁵⁄₁₆
28	16¼	29⁵⁄₁₆
29	16¹³⁄₁₆	30⅜
30	17⅜	31⁷⁄₁₆
31	17⅞	32½
32	18½	33½
33	19¹⁄₁₆	34⁹⁄₁₆
34	19¹¹⁄₁₆	35⅝
35	20¼	36¹¹⁄₁₆
36	20¹³⁄₁₆	37¾
37	21⅜	38¾
38	22	39¹³⁄₁₆
39	22½	40⅞
40	23⅛	41¹⁵⁄₁₆
41	23¹¹⁄₁₆	42¹⁵⁄₁₆
42	24¼	44
43	24⅞	45
44	25⁷⁄₁₆	46⅛
45	26	47⅛
46	26⅜	48³⁄₁₆
47	27⅛	49¼
48	27¾	50¼

Setback and Length of 60° Bends, Radius 1 to 72 Inches

Radius of Bend, Inches	Setback, Inches	Length of Bend, Inches
49	28$\frac{5}{16}$	51$\frac{5}{16}$
50	28$\frac{7}{8}$	52$\frac{3}{8}$
51	29$\frac{1}{2}$	53$\frac{7}{16}$
52	30	54$\frac{1}{2}$
53	30$\frac{5}{8}$	55$\frac{1}{2}$
54	31$\frac{3}{16}$	56$\frac{9}{16}$
55	31$\frac{3}{4}$	57$\frac{5}{8}$
56	32$\frac{3}{8}$	58$\frac{11}{16}$
57	32$\frac{15}{16}$	59$\frac{11}{16}$
58	33$\frac{1}{2}$	60$\frac{3}{4}$
59	34$\frac{1}{16}$	61$\frac{13}{16}$
60	34$\frac{5}{8}$	63
61	35$\frac{1}{4}$	64$\frac{1}{16}$
62	35$\frac{13}{16}$	65$\frac{1}{8}$
63	36$\frac{3}{8}$	66$\frac{1}{8}$
64	37	67$\frac{3}{16}$
65	37$\frac{9}{16}$	68$\frac{1}{4}$
66	38$\frac{1}{8}$	69$\frac{5}{16}$
67	38$\frac{11}{16}$	70$\frac{3}{8}$
68	39$\frac{1}{4}$	71$\frac{7}{16}$
69	39$\frac{7}{8}$	72$\frac{1}{2}$
70	40$\frac{3}{8}$	73$\frac{1}{2}$
71	41	74$\frac{9}{16}$
72	41$\frac{5}{8}$	75$\frac{5}{8}$

Setback and Length of 72° Bends

Setback and length of bend for 72° bends are calculated from the formulas:

$$\text{Setback} = \text{Radius} \times .727$$
$$= \text{Radius} \times 1.257$$

Calculated values for radius 1 to **72** inches are given in the tables beginning next page.

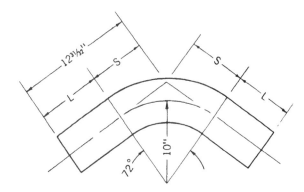

EXAMPLE

If a 24-inch length of pipe is to be used to construct a 72° bend of 10-inch radius with equal straight legs, what will be the length of legs?

SOLUTION

(1) In table on next page, for 10-inch radius:

Setback = 7¼ inches; Length of Bend = 12⅝₁₆ inches

(2) Length of two equal straight legs is equal to over-all length minus length of bend; therefore

$$2 \times L = 24 - 12\tfrac{5}{16} = 11\tfrac{7}{16}$$

(3) Length of each straight leg (L) = ½ × 11⅞₁₆ or 5²³⁄₃₂ inches.

Setback and Length of 72° Bends, Radius 1 to 72 Inches

Radius of Bend, Inches	Setback, Inches	Length of Bend, Inches
1	$\frac{3}{4}$	$1\frac{1}{4}$
2	$1\frac{7}{16}$	$2\frac{1}{2}$
3	$2\frac{3}{16}$	$3\frac{3}{4}$
4	$2\frac{15}{16}$	5
5	$3\frac{5}{8}$	$6\frac{5}{16}$
6	$4\frac{3}{8}$	$7\frac{9}{16}$
7	$5\frac{1}{16}$	$8\frac{13}{16}$
8	$5\frac{13}{16}$	$10\frac{1}{16}$
9	$6\frac{9}{16}$	$11\frac{5}{16}$
10	$7\frac{1}{4}$	$12\frac{9}{16}$
11	8	$13\frac{13}{16}$
12	$8\frac{3}{4}$	$15\frac{1}{16}$
13	$9\frac{7}{16}$	$16\frac{5}{16}$
14	$10\frac{3}{16}$	$17\frac{5}{8}$
15	$10\frac{7}{8}$	$18\frac{7}{8}$
16	$11\frac{5}{8}$	$20\frac{1}{8}$
17	$12\frac{3}{8}$	$21\frac{3}{8}$
18	$13\frac{1}{16}$	$22\frac{5}{8}$
19	$13\frac{13}{16}$	$23\frac{7}{8}$
20	$14\frac{9}{16}$	$25\frac{1}{8}$
21	$15\frac{1}{4}$	$26\frac{3}{8}$
22	16	$27\frac{5}{8}$
23	$16\frac{3}{4}$	$28\frac{15}{16}$
24	$17\frac{7}{16}$	$30\frac{3}{16}$

Setback and Length of 72° Bends, Radius 1 to 72 Inches

Radius of Bend, Inches	Setback, Inches	Length of Bend, Inches
25	18$\frac{3}{16}$	31$\frac{7}{16}$
26	18$\frac{7}{8}$	32$\frac{11}{16}$
27	19$\frac{5}{8}$	33$\frac{15}{16}$
28	20$\frac{3}{8}$	35$\frac{3}{16}$
29	21$\frac{1}{16}$	36$\frac{7}{16}$
30	21$\frac{13}{16}$	37$\frac{11}{16}$
31	22$\frac{9}{16}$	38$\frac{15}{16}$
32	23$\frac{1}{4}$	40$\frac{1}{4}$
33	24	41$\frac{1}{2}$
34	24$\frac{11}{16}$	42$\frac{3}{4}$
35	25$\frac{7}{16}$	44
36	26$\frac{3}{16}$	45$\frac{1}{4}$
37	26$\frac{7}{8}$	46$\frac{1}{2}$
38	27$\frac{5}{8}$	47$\frac{3}{4}$
39	28$\frac{3}{8}$	49
40	29$\frac{1}{16}$	50$\frac{1}{4}$
41	29$\frac{13}{16}$	51$\frac{9}{16}$
42	30$\frac{9}{16}$	52$\frac{13}{16}$
43	31$\frac{1}{4}$	54$\frac{1}{16}$
44	32	55$\frac{5}{16}$
45	32$\frac{11}{16}$	56$\frac{9}{16}$
46	33$\frac{7}{16}$	57$\frac{13}{16}$
47	34$\frac{3}{16}$	59$\frac{1}{16}$
48	34$\frac{7}{8}$	60$\frac{5}{16}$

Setback and Length of 72° Bends, Radius 1 to 72 Inches

Radius of Bend, Inches	Setback, Inches	Length of Bend, Inches
49	35⅝	61⁹⁄₁₆
50	36⅜	62⅞
51	37¹⁄₁₆	64⅛
52	37¹³⁄₁₆	65⅜
53	38⁹⁄₁₆	66⅝
54	39¼	67⅞
55	40	69⅛
56	40¹¹⁄₁₆	70⅜
57	41⁷⁄₁₆	71⅝
58	42³⁄₁₆	72¹⁵⁄₁₆
59	42⅞	74³⁄₁₆
60	43⅝	75⁷⁄₁₆
61	44⅜	76¹¹⁄₁₆
62	45¹⁄₁₆	77¹⁵⁄₁₆
63	45¹³⁄₁₆	79³⁄₁₆
64	46½	80⁷⁄₁₆
65	47¼	81¹¹⁄₁₆
66	48	82¹⁵⁄₁₆
67	48¹¹⁄₁₆	84¼
68	49⁷⁄₁₆	85½
69	50³⁄₁₆	86¾
70	50⅞	88
71	51⅝	89¼
72	52⅜	90½

Setback and Length of 90° Bends

Calculated values of setback and length of bend for 90° bends, radius 1 to 72 inches, are given in tables beginning next page. These values are based on the formulas:

$$\text{Setback} = \text{Radius} \times 1.0$$
$$\text{Length of Bend} = \text{Radius} \times 1.57$$

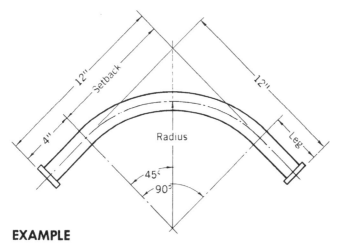

EXAMPLE

If at least 4 inches of straight pipe is required on each leg of a flanged 90° bend and the distance from the face of each flange perpendicular to the centerline of the opposite leg is 12 inches, what is the maximum length of radius possible in the bend? What will be the length of the flanged piece?

SOLUTION

(1) As in the sketch, 12 inches less the straight leg equals the setback. For a 4-inch leg, S = 12 − 4 = 8 inches.

(2) In the table, next page, for 8-inch setback:
Radius = 8 inches; Length of Bend = 12 %6 inches

(3) Length of piece is (4 × 2) + 12%6 = 20%6 inches.

Setback and Length of 90° Bends, Radius 1 to 72 Inches

Radius of Bend, Inches	Setback, Inches	Length of Bend, Inches
1	1	$1\frac{9}{16}$
2	2	$3\frac{1}{8}$
3	3	$4\frac{11}{16}$
4	4	$6\frac{5}{16}$
5	5	$7\frac{7}{8}$
6	6	$9\frac{7}{16}$
7	7	11
8	8	$12\frac{9}{16}$
9	9	$14\frac{1}{8}$
10	10	$15\frac{11}{16}$
11	11	$17\frac{1}{4}$
12	12	$18\frac{7}{8}$
13	13	$20\frac{7}{16}$
14	14	22
15	15	$23\frac{9}{16}$
16	16	$25\frac{1}{8}$
17	17	$26\frac{11}{16}$
18	18	$28\frac{1}{4}$
19	19	$29\frac{7}{8}$
20	20	$31\frac{7}{16}$
21	21	33
22	22	$34\frac{9}{16}$
23	23	$36\frac{1}{8}$
24	24	$37\frac{11}{16}$

Setback and Length of 90° Bends, Radius 1 to 72 Inches

Radius of Bend, Inches	Setback, Inches	Length of Bend, Inches
25	25	39¼
26	26	40¹³⁄₁₆
27	27	42⁷⁄₁₆
28	28	44
29	29	45⁹⁄₁₆
30	30	47⅛
31	31	48¹¹⁄₁₆
32	32	50¼
33	33	51¹³⁄₁₆
34	34	53⁷⁄₁₆
35	35	55
36	36	56⁹⁄₁₆
37	37	58⅛
38	38	59¾
39	39	61¼
40	40	62⅞
41	41	64⅜
42	42	66
43	43	67⁹⁄₁₆
44	44	69⅛
45	45	70¹¹⁄₁₆
46	46	72¼
47	47	73⅞
48	48	75⁷⁄₁₆

Setback and Length of 90° Bends, Radius 1 to 72 Inches

Radius of Bend, Inches	Setback, Inches	Length of Bend, Inches
49	49	77
50	50	78⅜₆
51	51	80⅛
52	52	81¹¹⁄₁₆
53	53	83¼
54	54	84⅞
55	55	86⅜
56	56	88
57	57	89½
58	58	91⅛
59	59	92¹¹⁄₁₆
60	60	94¼
61	61	95¹³⁄₁₆
62	62	97½
63	63	99
64	64	100½
65	65	102
66	66	103⅝
67	67	105⁷⁄₁₆
68	68	106¹⁵⁄₁₆
69	69	108⁷⁄₁₆
70	70	110
71	71	111½
72	72	113⅝₆

Setback and Length of 112½° Bends

Calculated values of setback and length of bend to the nearest ⅛ inch for 112½° bends, radius 1 to 48 inches, are given in tables beginning next page. These values are based on the formulas:

$$\text{Setback} = \text{Radius} \times 1.50$$
$$\text{Length of Bend} = \text{Radius} \times 1.96$$

Note that a 112½° bend equals a 90° bend plus a 22½° bend.

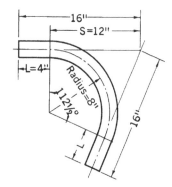

EXAMPLE

What is the length of pipe required for a piece including a 112½° bend on 8-inch radius such that the distance from each end of the piece to the point where centerlines of straight legs meet is 16 inches? How would you mark straight piece for bending?

SOLUTION

(1) In the table, next page, for 8-inch radius, Setback = 12 inches; Length of Bend = 15¾ inches

(2) Straight Leg = 16 − Setback = 4 inches

Length of Piece = 2 Straight Legs + Bend
$$= (2 \times 4) + 15\frac{3}{4} = 23\frac{3}{4} \text{ inches}$$

(3) Mark the beginning of the bend 4 inches from pipe end.

Mark end of bend 4 + 15¾ = 19¾ inches from end.

Setback and Length of 112½° Bends, Radius 1 to 48 Inches

Radius of Bend, Inches	Setback, Inches	Length of Bend, Inches
1	1½	2
2	3	3⅞
3	4½	5⅞
4	6	7⅞
5	7½	9⅞
6	9	11¾
7	10½	13¾
8	12	15¾
9	13½	17⅝
10	15	19⅝
11	16½	21⅝
12	18	23½
13	19½	25½
14	21	27½
15	22½	29½
16	24	31⅜
17	25½	33⅜
18	27	35⅜
19	28⅜	37¼
20	29⅞	39¼
21	31⅜	41¼
22	32⅞	43¼
23	34⅜	45⅛
24	35⅞	47⅛

Setback and Length of 112½° Bends, Radius 1 to 48 Inches

Radius of Bend, Inches	Setback, Inches	Length of Bend, Inches
25	37⅜	49⅛
26	38⅞	51
27	40⅜	53
28	41⅞	55
29	43⅜	57
30	44⅞	58⅞
31	45⅜	60⅞
32	47⅞	62⅞
33	49⅜	64¾
34	50⅞	66⅝
35	52⅜	68⅝
36	53⅞	70⅝
37	55⅜	72½
38	56⅞	74½
39	58⅜	76½
40	59⅞	78½
41	61⅜	80½
42	62⅞	82½
43	64⅜	83⅜
44	65⅞	86¼
45	67⅜	88¼
46	68⅞	90¼
47	70⅜	92¼
48	71⅞	94⅛

Setback and Length of 120° Bends

Values of setback and length of bend to the nearest ⅛ inch are given in tables beginning on the next page. These values are calculated from the formulas:

$$\text{Setback} = \text{Radius} \times 1.732$$
$$\text{Length of Bend} = \text{Radius} \times 2.09$$

Note that a 120° bend equals a 90° bend plus a 30° bend.

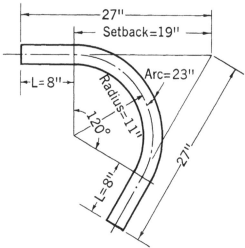

EXAMPLE

What will be the total length of pipe required for a piece including a 120° bend on 11-inch radius if the distance from either end to the point where centerlines of straight legs meet is 27 inches? How far from the end should the bend start?

SOLUTION

(1) In the table, next page, for 11-inch radius, Setback = 19 inches; Length of Bend = 23 inches

(2) Leg L = 27 − 19 = 8 inches
Length of Piece = (2 × 8) + 23 = 39 inches

(3) Start bend at length of leg = 8 inches from end.

Setback and Length of 120° Bends, Radius 1 to 48 Inches

Radius of Bend, Inches	Setback, Inches	Length of Bend, Inches
1	$1\frac{3}{4}$	$2\frac{1}{8}$
2	$3\frac{1}{2}$	$4\frac{1}{4}$
3	$5\frac{1}{4}$	$6\frac{1}{4}$
4	$6\frac{7}{8}$	$8\frac{3}{8}$
5	$8\frac{5}{8}$	$10\frac{1}{2}$
6	$10\frac{3}{8}$	$12\frac{5}{8}$
7	$12\frac{1}{8}$	$14\frac{5}{8}$
8	$13\frac{7}{8}$	$16\frac{3}{4}$
9	$15\frac{5}{8}$	$18\frac{7}{8}$
10	$17\frac{3}{8}$	21
11	19	23
12	$20\frac{3}{4}$	$25\frac{1}{8}$
13	$22\frac{1}{2}$	$27\frac{1}{4}$
14	$24\frac{1}{4}$	$29\frac{1}{4}$
15	26	$31\frac{1}{2}$
16	$27\frac{3}{4}$	$33\frac{1}{2}$
17	$29\frac{1}{2}$	$35\frac{5}{8}$
18	$31\frac{1}{8}$	$37\frac{3}{4}$
19	$32\frac{7}{8}$	$39\frac{3}{4}$
20	$34\frac{5}{8}$	$41\frac{7}{8}$
21	$36\frac{3}{8}$	44
22	$38\frac{1}{8}$	$46\frac{1}{8}$
23	$39\frac{7}{8}$	$48\frac{1}{4}$
24	$41\frac{5}{8}$	$50\frac{1}{4}$

Setback and Length of 120° Bends, Radius 1 to 48 Inches

Radius of Bend, Inches	Setback, Inches	Length of Bend, Inches
25	43¼	52⅜
26	45	54½
27	46¾	56½
28	47½	58⅝
29	50¼	60¾
30	52	62⅞
31	53¾	65⅛
32	55½	67¼
33	57⅛	69¼
34	58⅞	71⅜
35	60½	73½
36	62⅜	75⅝
37	64	77¾
38	65¾	79¾
39	67½	81⅞
40	69¼	84
41	71	86
42	72¾	88⅛
43	74½	90¼
44	76¼	92⅜
45	77⅞	94½
46	79⅝	96½
47	81⅜	98¾
48	83⅛	100¾

Setback and Length of 135° Bends

Values of setback and length of 135° bends to the nearest ⅛ inch are given in tables beginning on the next page. These values are calculated from the formulas:

$$\text{Setback} = \text{Radius} \times 2.41$$
$$\text{Length of Bend} = \text{Radius} \times 2.36$$

Note that a 135° bend equals a 90° bend plus a 45° bend.

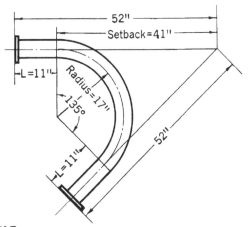

EXAMPLE

What will be the total length of a flanged pipe with 135° bend on 17-inch radius if the distance from either flange face to the point where centerlines of straight legs meet is 52 inches? How long will straight legs be, including flanges?

SOLUTION

(1) In the table, next page, for 17-inch radius,
Setback = 41 inches; Length of Bend = 40 inches

(2) Leg L = 52 − 41 = 11 inches

(3) Length of Pipe = 2L + Length of Bend
= (2 × 11) + 40 = 62 inches

Setback and Length of 135° Bends, Radius 1 to 48 Inches

Radius of Bend, Inches	Setback, Inches	Length of Bend, Inches
1	$2\frac{3}{8}$	$2\frac{3}{8}$
2	$4\frac{7}{8}$	$4\frac{3}{4}$
3	$7\frac{1}{4}$	$7\frac{1}{8}$
4	$9\frac{5}{8}$	$9\frac{3}{8}$
5	$12\frac{1}{8}$	$11\frac{3}{4}$
6	$14\frac{1}{2}$	$14\frac{1}{8}$
7	$16\frac{7}{8}$	$16\frac{1}{2}$
8	$19\frac{3}{8}$	$18\frac{7}{8}$
9	$21\frac{3}{4}$	$21\frac{1}{4}$
10	$24\frac{1}{8}$	$23\frac{1}{2}$
11	$26\frac{1}{2}$	$25\frac{7}{8}$
12	29	$28\frac{1}{4}$
13	$31\frac{3}{8}$	$30\frac{5}{8}$
14	$33\frac{3}{4}$	33
15	$36\frac{1}{4}$	$35\frac{3}{8}$
16	$38\frac{5}{8}$	$37\frac{3}{4}$
17	41	40
18	$43\frac{1}{2}$	$42\frac{3}{8}$
19	$45\frac{7}{8}$	$44\frac{3}{4}$
20	$48\frac{1}{4}$	$47\frac{1}{8}$
21	$50\frac{3}{4}$	$49\frac{1}{2}$
22	$53\frac{1}{8}$	$51\frac{7}{8}$
23	$55\frac{1}{2}$	$54\frac{1}{4}$
24	58	$56\frac{1}{2}$

Setback and Length of 135° Bends, Radius 1 to 48 Inches

Radius of Bend, Inches	Setback, Inches	Length of Bend, Inches
25	60⅜	59
26	62¾	61⅜
27	65⅛	63¾
28	67⅝	66⅛
29	70	68½
30	72⅜	70¾
31	74⅝	73¼
32	77⅛	75½
33	79½	77⅞
34	81⅞	80¼
35	84⅜	82⅝
36	86¾	85
37	89⅛	87¼
38	91½	89⅝
39	94	92
40	96⅜	94⅜
41	98¾	96¾
42	101¼	99⅛
43	103¾	101⅝
44	106⅛	103⅞
45	108½	106⅜
46	110⅞	108¾
47	113⅜	111
48	115¾	113⅜

Setback and Length of 150° Bends

Values of setback and length of 150° bends to the nearest $\frac{1}{8}$ inch are given in tables beginning on the next page. These values are calculated from the formulas:

$$\text{Setback} = \text{Radius} \times 3.73$$
$$\text{Length of Bend} = \text{Radius} \times 2.62$$

Note that a 150° bend equals a 90° bend plus a 60° bend.

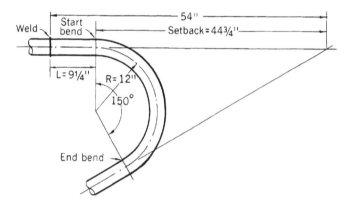

EXAMPLE

A piece of pipe with 150° bend on 12-inch radius is to be welded to a straight pipe so that the distance from the end of the straight pipe to the point where centerlines of bend legs meet is 54 inches. How far from the pipe end should the bend begin? What is the minimum length of pipe required to complete the bend?

SOLUTION

(1) In the table, next page, for 12-inch radius, Setback = 44 $\frac{3}{4}$ inches; Length of Bend = 31$\frac{3}{8}$ inches

(2) Start bend at 54 − 44$\frac{3}{4}$ = 9$\frac{1}{4}$ inches from end of pipe

(3) Minimum length of pipe to end of bend is leg L + Length of Bend = 9$\frac{1}{4}$ + 31$\frac{3}{8}$ = 40$\frac{5}{8}$ inches.

Setback and Length of 150° Bends, Radius 1 to 48 Inches

Radius of Bend, Inches	Setback, Inches	Length of Bend, Inches
1	3¾	2⅝
2	7½	5¼
3	11¼	7⅞
4	14⅞	10½
5	18⅝	13⅛
6	22⅜	15¾
7	26⅛	18⅜
8	29⅞	21
9	33⅝	23½
10	37⅜	26⅛
11	41	28¾
12	44¾	31⅜
13	48½	34
14	52¼	36⅝
15	56	39¼
16	59¾	41⅞
17	63½	44½
18	67⅛	47⅛
19	70⅞	49¾
20	74⅝	52⅜
21	78⅜	55
22	82⅛	57⅝
23	85⅞	60¼
24	89⅝	62⅞

Setback and Length of 150° Bends, Radius 1 to 48 Inches

Radius of Bend, Inches	Setback, Inches	Length of Bend, Inches
25	93¼	65½
26	97	68⅛
27	100¾	70⅝
28	104½	73¼
29	108¼	75⅞
30	112	78½
31	115¾	81¼
32	119½	83⅞
33	123¼	86½
34	127	89⅛
35	130¾	91¾
36	134⅜	94⅜
37	138⅛	97
38	141	99½
39	145½	102⅛
40	149¼	105
41	153	107⅜
42	156¾	110
43	160½	112¹³⁄₁₆
44	164¼	115³⁄₁₆
45	168	118
46	171¾	120⅝
47	175½	123⅛
48	179⅛	125¾

Setback and Length of 180° Bends

Setback for bends of 180° would be infinite if defined as a distance to the meeting point of straight leg center-lines, as was done at the beginning of this section. For practical purposes, the 180° bend is regarded as two 90° bends, in each of which the setback is equal to the radius.

Setback = Radius

Length of Bend = Radius × 3.14

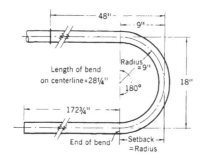

EXAMPLE

A 20-foot length of pipe is to be bent 180° on a 9-inch radius and welded to the end of another pipe so that the distance from the weld to a line drawn at the center of the bend, at right angle to the radius, is 48 inches. How far from the pipe end should the bend begin? How much straight pipe will be left at the other end?

SOLUTION

(1) Bend will start at 48 − 9 = 39 inches from end to be welded, since setback = radius.

(2) From table, next page, for radius or setback 9 inches,

Length of Bend = $28\frac{1}{4}$ inches

Length of straight pipe left at other end will be

(20 ft. × 12) − (39 + $28\frac{1}{4}$) = 240 − $67\frac{1}{4}$ = $172\frac{3}{4}$ inches

Setback and Length of 180° Bends, Radius 1 to 40 Inches

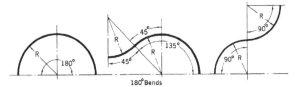

180° Bends

Radius and Setback, Inches	Length of Bend, Inches	Radius and Setback, Inches	Length of Bend, Inches
1	$3\frac{1}{8}$	21	66
2	$6\frac{1}{4}$	22	$69\frac{1}{8}$
3	$9\frac{3}{8}$	23	$72\frac{1}{4}$
4	$12\frac{5}{8}$	24	$75\frac{3}{8}$
5	$15\frac{3}{4}$	25	$78\frac{1}{2}$
6	$18\frac{7}{8}$	26	$81\frac{5}{8}$
7	22	27	$84\frac{7}{8}$
8	$25\frac{1}{8}$	28	88
9	$28\frac{1}{4}$	29	$91\frac{1}{8}$
10	$31\frac{3}{8}$	30	$94\frac{1}{4}$
11	$34\frac{1}{2}$	31	$97\frac{1}{2}$
12	$37\frac{3}{4}$	32	$100\frac{5}{8}$
13	$40\frac{7}{8}$	33	$103\frac{3}{4}$
14	44	34	$106\frac{3}{4}$
15	$47\frac{1}{8}$	35	$109\frac{7}{8}$
16	$50\frac{1}{4}$	36	113
17	$53\frac{3}{8}$	37	$116\frac{1}{8}$
18	$56\frac{1}{2}$	38	$119\frac{1}{4}$
19	$59\frac{3}{4}$	39	$122\frac{3}{8}$
20	$62\frac{7}{8}$	40	$125\frac{1}{2}$

Setback and Length of Bend, Any Radius, ¼° to 180°

From the tables which follow, setback and length of bend can be found for bends of any angle to the nearest ¼° and any radius.

To show how to use the tables, the following example is intentionally more complex than will ordinarily be found in the field.

The idea, in solving for angles with fractions of degrees (for instance 11¼°) is simply to add the table values for the fraction to the values for the whole degrees and then multiply by the radius.

EXAMPLE

There are sixty minutes (60′) in one degree.

Find the setback and length of a bend of 11° 15′ on a radius of 15½ inches.

SOLUTION

Fifteen minutes (15′) = ¹⁵⁄₆₀ = ¼ of a degree

(1) The table, next page, shows that to find the setback:

For an angle of 11°, multiply the radius by .0962
For an angle of ¼°, multiply the radius by .0023
Add these two values,   .0985
For an angle of 11¼°, multiply the radius by .0985.
Setback for 11¼° bend on 15½-inch radius, then, is
        .0985 × 15½ = 1.527 or 1½ inches

(2) The table shows also that to find the length of bend:

For an angle of 11°, multiply the radius by .1920
For an angle of ¼°, multiply the radius by .0044
Adding, for 11¼°, multiply the radius by .1964
Length of bend for 11¼° bend on 15½-inch radius is
.1964 × 15½ = 3.044 = 3¹⁄₁₆ (to nearest ¹⁄₁₆ inch).

Setback and Length of Bend,
Any Radius, ¼° to 180°

When Angle of Bend in Degrees Is	To Find Setback	To Find Length of Bend
	Multiply Radius by	
* ¼ (= 15′)	.0023	.0044
½ (= 30′)	.0045	.0087
¾ (= 45′)	.0068	.0131
1	.0087	.0175
2	.0175	.0349
3	.0261	.0524
4	.0349	.0698
5	.0436	.0873
6	.0524	.1047
7	.0611	.1222
8	.0699	.1396
9	.0787	.1571
10	.0875	.1745
11	.0962	.1920
12	.1051	.2094
13	.1139	.2269
14	.1228	.2443
15	.1316	.2618
16	.1405	.2793
17	.1494	.2967
18	.1584	.3142
19	.1673	.3316

* For use of fractions, see previous page.

Setback and Length of Bend,
Any Radius, ¼° to 180°

When Angle of Bend in Degrees Is	To Find Setback	To Find Length of Bend
	Multiply Radius by	
20	.1763	.3491
21	.1853	.3665
22	.1944	.3840
23	.2034	.4014
24	.2126	.4189
25	.2216	.4363
26	.2309	.4538
27	.2400	.4712
28	.2493	.4887
29	.2587	.5061
30	.2679	.5236
31	.2773	.5411
32	.2867	.5585
33	.2962	.5760
34	.3057	.5934
35	.3153	.6109
36	.3249	.6283
37	.3345	.6458
38	.3443	.6632
39	.3541	.6807
40	.3640	.6981
41	.3738	.7156
42	.3839	.7330

Setback and Length of Bend, Any Radius, ¼° to 180°

When Angle of Bend in Degrees Is	To Find Setback	To Find Length of Bend
	Multiply Radius by	
43	.3939	.7505
44	.4040	.7679
45	.4141	.7854
46	.4245	.8029
47	.4348	.8203
48	.4452	.8378
49	.4557	.8552
50	.4663	.8727
51	.4769	.8901
52	.4877	.9076
53	.4985	.9250
54	.5095	.9425
55	.5205	.9599
56	.5317	.9774
57	.5429	.9948
58	.5543	1.0123
59	.5657	1.0297
60	.5774	1.0472
61	.5890	1.0647
62	.6009	1.0821
63	.6128	1.0996
64	.6249	1.1170
65	.6370	1.1345

Setback and Length of Bend, Any Radius, ¼° to 180°

When Angle of Bend in Degrees Is	To Find Setback	To Find Length of Bend
	Multiply Radius by	
66	.6494	1.1519
67	.6618	1.1694
68	.6745	1.1868
69	.6872	1.2043
70	.7002	1.2217
71	.7132	1.2392
72	.7265	1.2566
73	.7399	1.2741
74	.7536	1.2915
75	.7673	1.3090
76	.7813	1.3265
77	.7954	1.3439
78	.8098	1.3614
79	.8243	1.3788
80	.8391	1.3963
81	.8540	1.4137
82	.8693	1.4312
83	.8847	1.4486
84	.9004	1.4661
85	.9163	1.4835
86	.9325	1.5010
87	.9484	1.5184
88	.9657	1.5359

Setback and Length of Bend,
Any Radius, ¼° to 180°

When Angle of Bend in Degrees Is	To Find Setback	To Find Length of Bend
	Multiply Radius by	
89	.9827	1.5533
90	1.0000	1.5708
91	1.0176	1.5883
92	1.0355	1.6057
93	1.0538	1.6232
94	1.0724	1.6406
95	1.0913	1.6581
96	1.1106	1.6765
97	1.1303	1.6930
98	1.1504	1.7104
99	1.1708	1.7279
100	1.1918	1.7453
101	1.2131	1.7628
102	1.2349	1.7802
103	1.2572	1.7977
104	1.2799	1.8151
105	1.3032	1.8326
106	1.3270	1.8500
107	1.3514	1.8675
108	1.3764	1.8849
109	1.4019	1.9024
110	1.4281	1.9198
111	1.4550	1.9373

Setback and Length of Bend, Any Radius, ¼° to 180°

When Angle of Bend in Degrees Is	To Find Setback	To Find Length of Bend
	Multiply Radius by	
112	1.4826	1.9547
113	1.5108	1.9722
114	1.5399	1.9896
115	1.5697	2.0071
116	1.6003	2.0246
117	1.6318	2.0420
118	1.6643	2.0595
119	1.6977	2.0769
120	1.7321	2.0944
121	1.7675	2.1118
122	1.8040	2.1293
123	1.8418	2.1467
124	1.8807	2.1642
125	1.9210	2.1816
126	1.9626	2.1991
127	2.0057	2.2165
128	2.0503	2.2340
129	2.0965	2.2514
130	2.1445	2.2689
131	2.1943	2.2864
132	2.2460	2.3038
133	2.2998	2.3213
134	2.3558	2.3387

Setback and Length of Bend, Any Radius, ¼° to 180°

When Angle of Bend in Degrees Is	To Find Setback	To Find Length of Bend
	Multiply Radius by	
135	2.4142	2.3562
136	2.4751	2.3736
137	2.5386	2.3911
138	2.6051	2.4085
139	2.6746	2.4260
140	2.7475	2.4424
141	2.8239	2.4609
142	2.9042	2.4783
143	2.9887	2.4958
144	3.0777	2.5132
145	3.1716	2.5307
146	3.2708	2.5482
147	3.3759	2.5656
148	3.4874	2.5831
149	3.6059	2.6005
150	3.7320	2.6180
151	3.8667	2.6354
152	4.0108	2.6529
153	4.1653	2.6703
154	4.3315	2.6878
155	4.5170	2.7052
156	4.7046	2.7227
157	4.9151	2.7401

Setback and Length of Bend, Any Radius, ¼° to 180°

When Angle of Bend in Degrees Is	To Find Setback	To Find Length of Bend
	Multiply Radius by	
158	5.1445	2.7576
159	5.3955	2.7750
160	5.6713	2.7925
161	5.9758	2.8100
162	6.3137	2.8274
163	6.6911	2.8449
164	7.1154	2.8623
165	7.5957	2.8798
166	8.1443	2.8972
167	8.7769	2.9147
168	9.5144	2.9321
169	10.385	2.9496
170	11.430	2.9670
171	12.706	2.9845
172	14.301	3.0019
173	16.350	3.0194
174	19.081	3.0368
175	22.904	3.0543
176	28.636	3.0718
177	38.188	3.0892
178	57.290	3.1067
179	114.59	3.1241
180	1	3.1416

Offset Bends

A simple example of the 45° offset bend is illustrated and the method of fabrication outlined below. Data for other standard bends are given on the following pages. Explanations and formulas for bends at any angle are to be found in the section on trigonometry which appears in the latter part of this book.

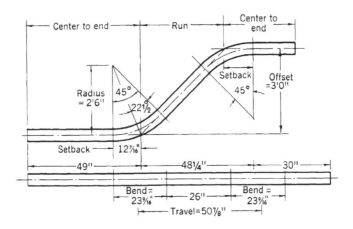

EXAMPLE

Lay out and make a 45° offset bend on 30-inch radius and with 36-inch offset so that the distance from the end of the pipe to the point where centerlines of straight legs of the first bend meet is 49 inches and the similar distance from the second bend to the other end of the pipe is 30 inches.

SOLUTION

(1) In tables for setback and length of 45° bends, find, for 30-inch radius:

$$\text{Setback} = 12\frac{7}{16} \text{ inches}$$
$$\text{Length of Bend} = 23\frac{9}{16} \text{ inches}$$

Offset Bends
(Continued)

(2) In section which follows, for 45° offset bend, find: Run = Offset; Travel = Offset × 1.414. Since the Offset is 36 inches, Travel = 36 × 1.414 = 50⅞ inches.

(3) From a point 49 inches from the end of the straight pipe, measure back (Setback =) 12⁷⁄₁₆ inches to locate beginning of first bend. Mark this point.

(4) From the marked beginning of the bend, measure forward (length of bend =) 23⁹⁄₁₆ inches and mark the end of the first bend.

(5) Notice in the sketch, opposite page, that the length of straight pipe between bends is equal to the travel minus twice the setback. Therefore, from the end of the first bend measure forward

$$50\tfrac{7}{8} - (2 \times 12\tfrac{7}{16}) = 26 \text{ inches}$$

This marks the beginning of the second bend.

(6) Since bends are equal, measure forward from the beginning of the second bend (length of bend =) 23⁹⁄₁₆ inches to mark the end of the second bend.

(7) Length of straight pipe at the end will be 30 inches minus the setback. Therefore, measure forward from the end of the second bend:

$$30 - 12\tfrac{7}{16} = 17\tfrac{9}{16} \text{ inches}$$

This marks the end of the pipe.

(8) Total length of pipe in the piece will be, reading along the straight pipe in the sketch:

$$49 - 12\tfrac{7}{16} + 23\tfrac{9}{16} + 26 + 23\tfrac{9}{16} + 17\tfrac{9}{16} = 127\tfrac{1}{4} \text{ inches}$$

(9) Before cutting off pipe or bending, make allowance for any joints or fittings as shown in the section on that subject.

15° Offset Bends

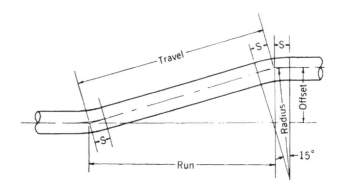

$$Offset = Run \times .2679$$
$$Run = Offset \times 3.7320$$
$$Travel = Offset \times 3.8637$$
$$Setback = Radius \times .132$$
$$Length\ of\ Bend = Radius \times .262$$

Calculated values of setback and length of 15° bends were given previously. Values for travel and run are on page opposite.

EXAMPLE

A pipe is offset 23½ inches using 15° bends. Find the travel and run.

SOLUTION

For values of travel and run other than those in table on opposite page, simply add component lengths. Thus, travel and run for 23½-inch offset would be:

	Travel	Run
For 12-inch offset	46⅜	44¾
For 11-inch offset	42½	41¹⁄₁₆
For ½-inch offset	1¹⁵⁄₁₆	1⅞
Total (for 23½-inch offset)	90¹³⁄₁₆	87¹¹⁄₁₆

15° Offset Bends, Two or More Pipes, Equal Spread

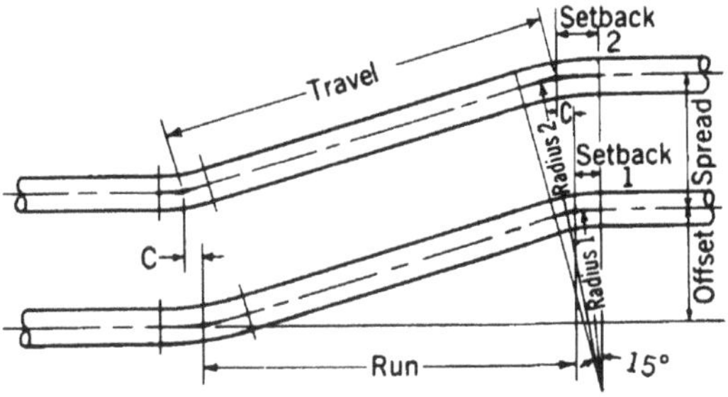

C = Spread × .132

Travel and Run		
Offset, Inches	**Travel, Inches**	**Run, Inches**
1/4	1	15/16
1/2	1 15/16	1 7/8
3/4	2 7/8	2 13/16
1	3 7/8	3 3/4
2	7 3/4	7 1/2
3	11 9/16	11 3/16
4	15 7/16	14 15/16
5	19 5/16	18 11/16
6	23 3/16	22 3/8
7	27 1/16	26 1/8
8	30 15/16	29 7/8
9	34 3/4	33 9/16
10	38 5/8	37 5/16
11	42 1/2	41 1/16
12	46 3/8	44 3/4

22½° Offset Bends

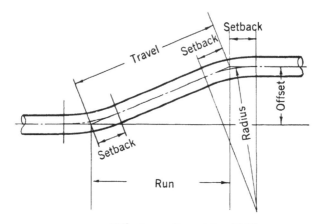

$$\text{Offset} = \text{Run} \times .4142$$
$$\text{Run} = \text{Offset} \times 2.4142$$
$$\text{Travel} = \text{Offset} \times 2.6131$$
$$\text{Setback} = \text{Radius} \times .199$$
$$\text{Length of Bend} = \text{Radius} \times .393$$

Calculated values of setback and length of 22½° bends were given previously. Values for travel and run are on page opposite.

EXAMPLE

A pipe is offset 2 ft. 6½ inches using 22½° bends. Find the travel and run.

SOLUTION

For values of travel and run other than those tabulated opposite, simply add component lengths. Thus, travel and run for 2 ft. 6½ inch offset would be:

	Travel	Run
For 12-inch offset	31⅜	28⅞
For 12-inch offset	31⅜	28⅞
For 6-inch offset	15¹¹⁄₁₆	14¼
For ½-inch offset	1⁵⁄₁₆	1³⁄₁₆
Total (for 2 ft. 6½ inches)	79¾	73³⁄₁₆

22½° Offset Bends,
Two or More Pipes, Equal Spread

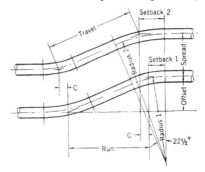

C = Spread × .1989

Travel and Run		
Offset, Inches	Travel, Inches	Run, Inches
¼	⅝	⅝
½	1⁵⁄₁₆	1³⁄₁₆
¾	1¹⁵⁄₁₆	1¹³⁄₁₆
1	2⅝	2⁷⁄₁₆
2	5¼	4¹³⁄₁₆
3	7⅞	7¼
4	10⁷⁄₁₆	9⅝
5	13¹⁄₁₆	12¹⁄₁₆
6	15¹¹⁄₁₆	14¼
7	18⅝	16⅞
8	20⅞	19⁵⁄₁₆
9	23½	21¹¹⁄₁₆
10	26⅛	24⅛
11	28¾	26⁷⁄₁₆
12	31⅜	28⅞

30° Offset Bends

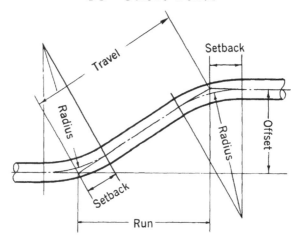

$$\text{Offset} = \text{Run} \times .5774$$
$$\text{Run} = \text{Offset} \times 1.7321$$
$$\text{Travel} = \text{Offset} \times 2.0000$$
$$\text{Setback} = \text{Radius} \times .268$$
$$\text{Length of Bend} = \text{Radius} \times .524$$

Calculated values of setback and length of 30° bends were given previously. Values for travel and run are on page opposite. For offsets not shown in the table, simply add together values for given offsets totaling the desired amount.

EXAMPLE
A pipe is offset 22¾ inches using 30° bends. Find the travel and run.

SOLUTION
Adding values from the table, opposite page:

	Travel	Run
For 12-inch offset	24	20¹³⁄₁₆
For 10-inch offset	20	17⁵⁄₁₆
For ¾-inch offset	1½	1⁵⁄₁₆
Total (for 22¾ inch offset)	45½	39⁷⁄₁₆

30° Offset Bends, Two or More Pipes, Equal Spread

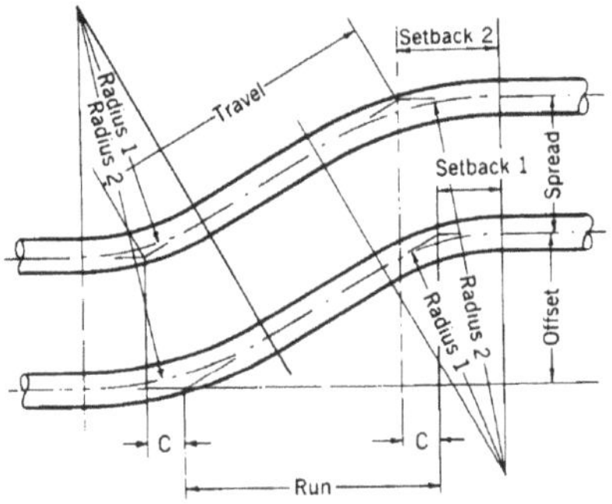

C = Spread × .2679

Travel and Run		
Offset, Inches	**Travel, Inches**	**Run, Inches**
1/4	1/2	7/16
1/2	1	7/8
3/4	1 1/2	1 5/16
1	2	1 3/4
2	4	3 5/16
3	6	5 3/16
4	8	6 15/16
5	10	8 5/8
6	12	10 1/2
7	14	12 1/8
8	16	13 7/8
9	18	15 9/16
10	20	17 5/16
11	22	19
12	24	20 13/16

45° Offset Bends

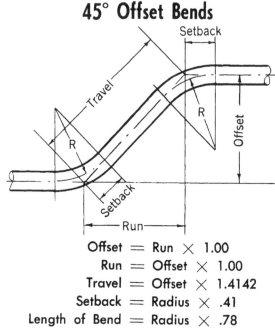

$$\text{Offset} = \text{Run} \times 1.00$$
$$\text{Run} = \text{Offset} \times 1.00$$
$$\text{Travel} = \text{Offset} \times 1.4142$$
$$\text{Setback} = \text{Radius} \times .41$$
$$\text{Length of Bend} = \text{Radius} \times .78$$

Calculated values of setback and length of 45° bends were given previously. Values for travel and run are on page opposite. For offsets not covered by the table, add together given offsets totaling the desired amount.

EXAMPLE

Find travel and run in a pipe offset $18\frac{1}{4}$ inches using 45° bends.

SOLUTION

As stated in formulas above, run = offset = $18\frac{1}{4}$ inches.

Adding values for travel from table, opposite page:

For 12-inch offset, travel	= 17	inches
For 6-inch offset, travel	= $8\frac{1}{2}$	
For $\frac{1}{4}$-inch offset, travel	= $\frac{3}{8}$	

Total (for $18\frac{1}{4}$-inch offset)	= $25\frac{7}{8}$	inches

45° Offset Bends,
Two or More Pipes, Equal Spread

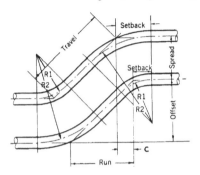

$$C = Spread \times .4142$$

Travel and Run		
Offset, Inches	**Travel, Inches**	**Run, Inches**
¼	⅜	¼
½	¾	½
¾	1¹⁄₁₆	¾
1	1⁷⁄₁₆	1
2	2¹³⁄₁₆	2
3	4¼	3
4	5⅝	4
5	7¹⁄₁₆	5
6	8½	6
7	9⅞	7
8	11⁵⁄₁₆	8
9	12¾	9
10	14⅛	10
11	15⁹⁄₁₆	11
12	17	12

60° Offset Bends

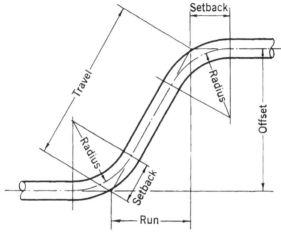

$$Offset = Run \times 1.7320$$
$$Run = Offset \times .5774$$
$$Travel = Offset \times 1.1547$$
$$Setback = Radius \times .58$$
$$Length \ of \ Bend = Radius \times 1.05$$

Calculated values of setback and length of 60° bends were given previously. Values for travel and run are on page opposite. For values not shown in the table, simply add values for given offsets totaling the desired amount.

EXAMPLE

A pipe is offset 16½ inches using 60° bends. Find the travel and run.

SOLUTION

Adding values from the table opposite:

	Travel	Run
For 12-inch offset	13⅞ Inches	6¹⁵⁄₁₆ Inches
For 4-inch offset	4⅝	2⁵⁄₁₆
For ½-inch offset	⁹⁄₁₆	⁵⁄₁₆
Total (for 16½-inch offset)	19¹⁄₁₆ Inches	9⁹⁄₁₆ Inches

60° Offset Bends,
Two or More Pipes, Equal Spread

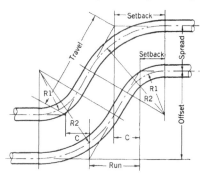

$$C = Spread \times .5774$$

Travel and Run		
Offset, Inches	Travel, Inches	Run, Inches
$\frac{1}{4}$	$\frac{5}{16}$	$\frac{1}{8}$
$\frac{1}{2}$	$\frac{9}{16}$	$\frac{5}{16}$
$\frac{3}{4}$	$\frac{7}{8}$	$\frac{7}{16}$
1	$1\frac{1}{8}$	$\frac{9}{16}$
2	$2\frac{5}{16}$	$1\frac{1}{8}$
3	$3\frac{7}{16}$	$1\frac{3}{4}$
4	$4\frac{5}{8}$	$2\frac{5}{16}$
5	$5\frac{3}{4}$	$2\frac{7}{8}$
6	$6\frac{15}{16}$	$3\frac{7}{16}$
7	$8\frac{1}{16}$	$4\frac{1}{16}$
8	$9\frac{1}{4}$	$4\frac{5}{8}$
9	$10\frac{3}{8}$	$5\frac{3}{16}$
10	$11\frac{9}{16}$	$5\frac{3}{4}$
11	$12\frac{11}{16}$	$6\frac{3}{8}$
12	$13\frac{7}{8}$	$6\frac{15}{16}$

72° Offset Bends

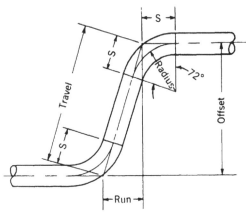

Offset = Run   × 3.0777
Run = Offset × .3249
Travel = Offset × 1.0515
Setback = Radius × .727
Length of Bend = Radius × 1.257

Calculated values of setback and length of 72° bends were given previously. Values for travel and run are on page opposite. For values not shown in the table, simply add values for given offsets totaling the desired amount.

EXAMPLE

Find travel and run in a pipe which is offset 21½ inches using 72° bends.

SOLUTION

From table on opposite page, add values of travel and run for component lengths.

	Travel	Run
For 11-inch offset	11⁹⁄₁₆	3⁹⁄₁₆
For 10-inch offset	10½	3¼
For ½-inch offset	½	³⁄₁₆
Total (for 21½-inch offset)	22⁹⁄₁₆	7

72° Offset Bends,
Two or More Pipes, Equal Spread

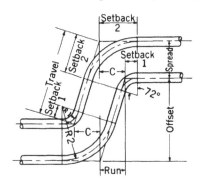

C = Spread × .727

Travel and Run		
Offset, Inches	Travel, Inches	Run, Inches
¼	¼	⅟₁₆
½	½	³⁄₁₆
¾	¹³⁄₁₆	¼
1	1⅟₁₆	⁵⁄₁₆
2	2⅛	⅝
3	3³⁄₁₆	1
4	4³⁄₁₆	1⁵⁄₁₆
5	5¼	1⅝
6	6⁵⁄₁₆	1¹⁵⁄₁₆
7	7⅜	2¼
8	8⁷⁄₁₆	2⅝
9	9½	2¹⁵⁄₁₆
10	10½	3¼
11	11⁹⁄₁₆	3⁹⁄₁₆
12	12⅝	3⅞

90° Offset Bends
One or More Pipes, Equal Spread

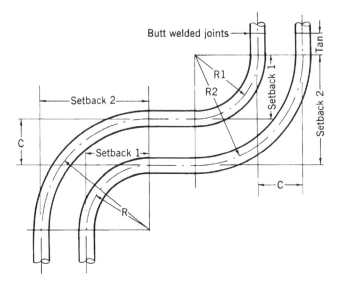

Offset = any amount greater than
Radius × 2
Run = 0
Travel = Offset
C = Spread
Setback = Radius
Length of Bend = Radius × 1.57

Travel in a 90° offset falls along and is equal to the offset. The drawing shows that the shortest possible offset is equal to twice the shortest radius to which pipe can be bent. Shortest radii for standard pipe are shown in the latter part of this book.

The drawing illustrates the fact that a certain amount of straight pipe (marked Tan for tangent) must be left at the connection end of each bend. Values for minimum length of tangent are also shown in the section on pipe standards.

Continuous Bend Offsets

Where offsets are made by two equal and opposite bends with no straight pipe between bends, the amounts of offset, end-to-end run, and length of pipe are fixed by the radius and angle of bending.

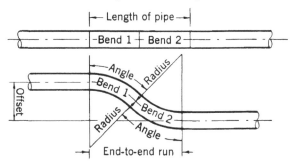

When Offset Bend Is Made by:							
Two 30° Bends				Two 45° Bends			
Radius, Inches	Offset, Inches	E to E Run, Inches	Length of Pipe, Inches	Radius, Inches	Offset, Inches	E to E Run, Inches	Length of Pipe, Inches
2	½	2	2⅛	2	1³⁄₁₆	2¹³⁄₁₆	3⅛
4	1	4	4³⁄₁₆	4	2⅜	5⅝	6⁵⁄₁₆
6	1½	6	6⁵⁄₁₆	6	3½	8½	9⁷⁄₁₆
8	2	8	8¼	8	4¹¹⁄₁₆	11⁵⁄₁₆	12⁹⁄₁₆
10	2½	10	10⁷⁄₁₆	10	5⅞	14⅛	15¹¹⁄₁₆
12	3	12	12⁹⁄₁₆	12	7	17	18⅞

For all angles and radii, it can be shown by trigonometry that for continuous bend offsets:

$$\text{Offset} = 2 \times \text{Radius} \times (1 - \text{Sine of Angle})$$

$$\text{End-to-End Run} = 2 \times \text{Radius} \times \text{Sine of Angle}$$

$$\text{Length of Pipe} = .035 \times \text{Radius} \times \text{No. of Degrees in Angle}$$

Double Offset Bends

An example of a 60° double offset bend is illustrated and the method of fabrication outlined below. Solution is merely extension of data already calculated for 60° bends and 60° offset bends.

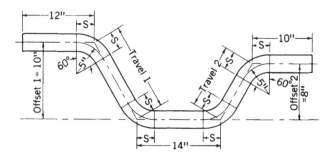

EXAMPLE

Lay out and make a 60° double offset bend on 5-inch radii so that the first offset is 10 inches and return offset is 8 inches. Let the distance from the end of pipe to point where centerlines of straight legs at the first bend meet be 12 inches, and the similar distance from the other end of pipe to the fourth bend be 10 inches. Let the distance along centerline of the parallel offset length between points of intersection of centerlines of straight legs which form the double offset be 14 inches.

SOLUTION

(1) In tables for setback and length of 60° bends, find, for 5-inch radius: Setback = 2⅞ inches; Length of Bend = 5¼ inches.

(2) In section on 60° offset bends, find: Run = offset × .5774; Travel = Offset × 1.1547. For first offset of 10 inches, Run = 10 × .5774 = 5¾ inches; Travel = 10 × 1.1547 = 11%₆ inches. For return offset of 8 inches, Run = 8 × .5774 = 4⅝ inches; Travel = 8 × 1.1547 = 9¼ inches.

Double Offset Bends

(Continued)

(3) From a point 12 inches from the end of the straight pipe, measure back (setback $=$) $2\frac{7}{8}$ inches to locate beginning of first bend. Mark this point.

(4) From the marked beginning of the bend, measure forward (length of bend $=$) $5\frac{1}{4}$ inches and mark the end of first bend.

(5) Length of straight pipe between bends is equal to the travel minus twice the setback. From end of first bend measure forward: $11\frac{9}{16} - (2 \times \frac{7}{8}) = 5\frac{13}{16}$

(6) Since bends are equal, measure forward from the beginning of the second bend (length of bend $=$) $5\frac{1}{4}$ inches to mark the end of second bend.

(7) Length of straight pipe in the parallel offset section will be 14 inches minus the setback of the second and third bends. Therefore, measure forward from the end of the second bend: $14 - (2 \times 2\frac{7}{8}) = 8\frac{1}{4}$

(8) Measure forward from beginning of third bend the length of bend equal to $5\frac{1}{4}$ inches, as in the case of first and second bends, to mark end of third bend.

(9) Length of straight pipe between bends three and four is equal to travel minus twice the setback. Therefore, from the end of third bend measure forward:

$$9\frac{1}{4} - (2 \times 2\frac{7}{8}) = 3\frac{1}{2}$$

(10) Since bends are equal, measure forward from the beginning of fourth bend $5\frac{1}{4}$ inches to mark the end of the fourth bend.

(11) Length of straight pipe at the end will be 10 inches minus the setback. Therefore, measure forward from end of fourth bend: $10 - 2\frac{7}{8} = 7\frac{1}{8}$ inches

(12) Total length of pipe in the double offset piece:
$12 - 2\frac{7}{8}. + 5\frac{1}{4} + 5\frac{13}{16} + 5\frac{1}{4} + 8\frac{1}{4} + 5\frac{1}{4} + 3\frac{1}{2}$
$+ 5\frac{1}{4} + 7\frac{1}{8} = 54\frac{13}{16}$ inches total length

Crossover Bend 180°

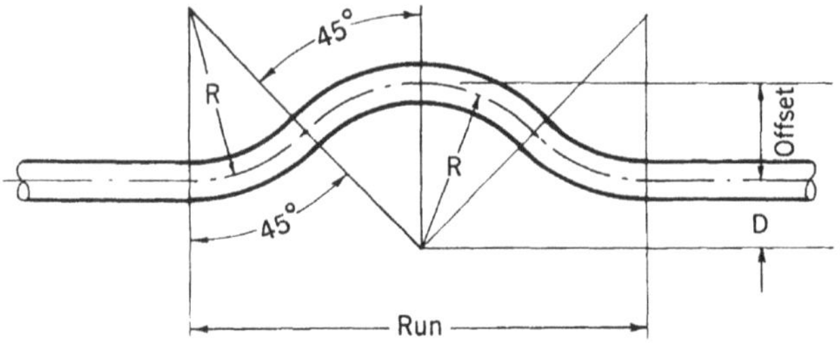

The 180° crossover bend, as shown above, is a continuous series of four 45° bends totaling 180°. If the radius is known, other dimensions are as follows:

Offset = Radius × .5858
Run = Radius × 2.828
D = Radius × .4142 = Radius − Offset

Length of Pipe in Bend = Radius × 3.1416

For cases where offset is known, and radius is desired:
Radius = Offset × 1.707

Calculated Values			
Radius, Inches	**Offset, Inches**	**Run, Inches**	**Length of Pipe in Bend, Inches**
4	2⅜	11⅜	12 9/16
6	3½	17	18⅞
8	4 11/16	22 9/16	25⅛
10	5⅞	28¼	31 7/16
12	7	33⅞	37¾
14	8 3/16	39½	44
16	9⅜	45 7/16	50¼
18	10½	50⅞	56⅝

Other 180° Bends

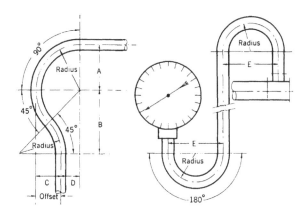

At left above is a single offset quarter bend of 180°. At right above is a steam gage siphon with two 180° U-bends.

Length of Pipe in 180° Bends = Radius × 3.1416

For cases where offset is known and radius is desired:
Radius = Offset (C) × 1.707

Dimensions as marked in drawings are:

    A = Radius
    B = Radius × 1.4142
    C = Radius × .5858
    D = Radius × .4142
    E = Radius × 2

Notice that the 180° bend at left above is made up of two 45° bends and one 90° bend, while the bends in the right hand illustration are continuous arcs of 180° each. In either case, the length of pipe in 180° of bend is the same for a given radius. That is, no matter what the direction of bending may be, if the radius is constant, the length of pipe in a total of 180° is equal to the radius times 3.1416.

Single Offset Quarter Bend – 210°

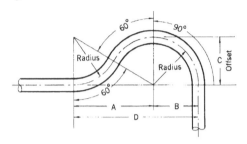

A = Radius × 1.7320
B = Radius
C = Radius = Offset
D = Radius × 2.7320
Length of Bend = Radius × 3.6651

Calculated Values					
Radius of Bend, Inches	Length of Bend, Inches	Dimension D, Inches	Radius of Bend, Inches	Length of Bend, Inches	Dimension D, Inches
1	3⅝	2¾	18	66	49¼
2	7⅜	5½	24	88	65½
3	11	8¼	36	132	98¼
4	14¾	11	48	176	131¼
5	18¼	13¾	60	220	164
6	22	16½	72	264	196½
7	25¾	19¼	84	308	229½
8	29¼	22	96	351¾	262
9	33	24½	108	395¾	295
10	36¾	27¼	120	439¾	328
11	40¼	30	132	483¾	360½
12	44	32¾	144	527¾	393½

Crossover Bend — 240°

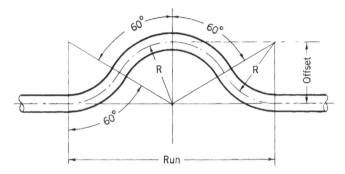

The 240° crossover bend is a series of four 60° bends, as shown. If the radius is known, other dimensions are as follows:

$$Offset = Radius$$
$$Run = Radius \times 3.4640$$
$$Length\ of\ Pipe\ in\ Bend = Radius \times 4.1887$$

Calculated Values		
Radius, Inches	Run, Inches	Length of Pipe, in Bend, Inches
4	13⅞	16¾
6	20¾	25⅛
8	27¾	33½
10	34⅝	41⅞
12	41⅝	50¼
14	48½	58⅝
16	55½	67
18	62⅜	75½
20	69¼	83¾
22	76¼	92⅛
24	83⅛	100½

Single Offset U-Bend—270°

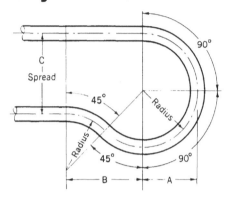

A = Bend Radius
B = Radius × 1.4142
C = Radius × 1.4142
Length of Bend = Radius × 4.7123

If the spread is known and it is desired to find the radius:

Radius = Spread (C) × .707

Calculated Values					
Radius, Inches	Length of Bend, Inches	B or C, Inches	Radius, Inches	Length of Bend, Inches	B or C, Inches
2	9½	2¾	12	56½	16⅞
3	14¼	4¼	18	84¾	25½
4	18¾	5⅝	24	113	34
5	23½	7⅛	36	169½	50¾
6	28¼	8½	48	226¼	68
7	33	9⅞	60	282¾	85
8	37¾	11¼	72	339¼	100¼
9	42½	12¾	84	395¾	118½
10	47¼	14⅛	96	452½	135¾
11	51¾	15½	120	565½	170

Single Offset U-Bend – 300°

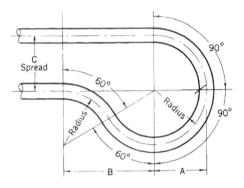

A = Bend Radius
B = Radius × 1.7320
C = Radius
Length of Bend = Radius × 5.2359

If the spread is known and it is desired to find the radius:

Radius = Spread (C)

Calculated Values					
Radius, Inches	Length of Bend, Inches	B, Inches	Radius, Inches	Length of Bend, Inches	B, Inches
2	10½	3½	12	62¾	20¾
3	15¾	5¼	18	94¼	31¼
4	21	6¾	24	125¾	41½
5	26¼	8¾	36	188½	62½
6	31½	10½	48	251¼	83¼
7	36¾	12¼	60	314¼	103¾
8	42	13¾	72	377	124½
9	47¼	15½	84	439¾	141¼
10	52½	17¼	96	502¾	166
11	57½	19	120	754	207½

Circular Bends — 360°

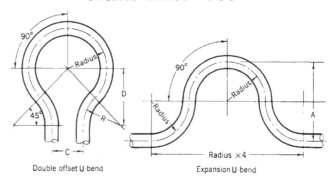

Double offset U-bend   Expansion U-bend

$$A = B = \text{Radius} \times 2$$
$$C = \text{Radius} \times .8284$$
$$\text{Radius} = C \times 1.207$$
$$D = \text{Radius} \times 1.4142$$
$$\text{Length of Bend} = \text{Radius} \times 6.2832$$

Calculated Values

Radius, Inches	C, Inches	Length of Bend, Inches	Radius, Inches	C, Inches	Length of Bend, Inches
2	1⅝	12½	12	9⅞	75⅜
3	2½	18⅞	18	10¾	113⅛
4	3¼	25⅛	24	19¾	150⅞
5	4⅛	31⅜	36	29	226¼
6	5	37⅝	48	39¾	301¾
7	5¾	44	60	49¾	377⅛
8	6⅝	50¼	72	59⅝	452½
9	7½	56½	84	69½	527¾
10	8¼	62⅞	96	79½	603
11	9⅛	69	120	99⅜	754

For calculated values of D, see dimension B for 270°
U-bend as given previously.

Expansion U-Bend — 540°

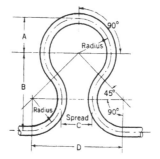

A = Radius
B = Radius × 2.4142
C = Radius × .8284
D = Radius × 2.8284
Length of Bend = Radius × 9.4246

If the spread is known and it is desired to find the radius:

$$\text{Radius} = C \times 1.207$$

Calculated Values				
Radius, Inches	B, Inches	C, Inches	D, Inches	Length of Bend, Inches
6	14½	5	17	56⅝
7	16⅞	5¾	19¾	66
8	19¼	6⅝	22⅝	75⅜
9	21¾	7½	25½	84¾
10	24⅛	8¼	28¼	94¼
12	28⅞	10	34	113⅛
18	43⅜	14⅞	50⅞	169¾
24	57⅞	19⅞	62⅞	226⅛
36	81¾	29¾	101⅞	339¼
48	115⅞	39¾	135¾	452½

Fabrication of Bends

In order to fabricate the 540° double offset expansion U-bend shown on the previous page, a certain order of bending should be followed as follows:

(1) Space and mark off the length of pipe required in each of the 90° and 45° component bends.

(2) Make the 45° offset bends.

(3) Make the two end 90° bends.

(4) Make the two 90° bends in the center, and the full 540° bend comes into position.

In preparation for any bend, and before actually bending the pipe, allowance must be made for straight ends (tangents), threaded ends, flanges, welding joints, and other clearances before cutting the pipe off at the overall length.

If heating by torch, start the first heat of the bend at the setback point with the torch moving forward. Heat pipe evenly with repeated passes toward the end of the bend length. As each heat is applied, form the bend with a bending machine or template.

Bending should be done so that a minimum of flattening takes place at the outside of the bend. Where possible, short sections of the bend should be formed continuously from one end of the arc to the other, rather than applying all bending pressure to the center of the arc.

Calculation for forming so-called wrinkle bends is shown in a later section, as is also the minimum bending radius for standard pipe.

The 90° Turn
One Pipe, Two-45° Bends

A 90° turn made to a given radius by two 45° pipe bends of a shorter radius has, bounding the bends, three straight lengths. These lengths plus the adjacent setbacks create distances having a definite relationship to each other and to the radius of the turn.

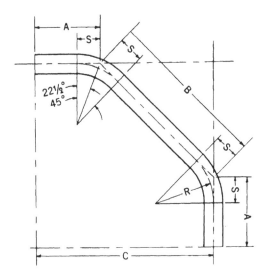

A = Radius of Turn (C) × .414
B = 2A = Radius of Turn (C) × .828

Since the Setback for 45° bends is equal to the Radius of Bend × .414, it is seen that the Tables for Setback and Length of 45° Bends may be used to find the value of A and 2A (B), using as radius, in this case, the Radius of Turn.

EXAMPLE

Find the length of pipe required to make a 90° turn of 24-inch radius with two 45° bends of 12-inch radius.

The 90° Turn
One Pipe, Two-45° Bends
(Continued)

SOLUTION

(1) From Tables for Setback and Length of 45° Bends, find for bends of 12-inch radius:

$$\text{Setback} = 4\tfrac{15}{16} \text{ inches}$$
$$\text{Length of Bend} = 9\tfrac{7}{16} \text{ inches}$$

(2) Using the setback column in the same table, find for radius of turn equal to 24 inches:

$$A = 9\tfrac{15}{16} \text{ inches}$$
$$B = 2A = 19\tfrac{7}{8} \text{ inches}$$

(3) From a point $9\tfrac{15}{16}$ inches from end of pipe, measure back (Setback =) $4\tfrac{15}{16}$ inches. This marks the beginning of first bend.

(4) From the beginning of the first bend, measure forward (Length of Bend =) $9\tfrac{7}{16}$ inches. This marks the end of first bend.

(5) Length of straight pipe between bends is equal to length B minus twice the setback. Therefore, from end of the first bend measure forward:

$$19\tfrac{7}{8} - (2 \times 4\tfrac{15}{16}) = 10 \text{ inches}$$

This marks the beginning of second bend.

(6) From beginning of second bend, measure forward $9\tfrac{7}{16}$ inches. This marks the end of second bend.

(7) Length of straight pipe at the end will be length A minus the setback. Therefore, measure forward from the end of second bend:

$$9\tfrac{15}{16} - 4\tfrac{15}{16} = 5 \text{ inches}$$

(8) Total length of pipe required, therefore, is:

$$9\tfrac{15}{16} - 4\tfrac{15}{16} + 9\tfrac{7}{16} + 10 + 9\tfrac{7}{16} + 5 = 38\tfrac{7}{8} \text{ inches}$$

(9) Before cutting off pipe or bending, make allowance for any joints or fittings as shown in the section on that subject.

The 90° Turn
One Pipe, Three-30° Bends

A 90° turn made to a given radius by three 30° pipe bends of shorter radius has, bounding the bends, four straight lengths. These lengths plus the adjacent setbacks create distances having a definite relationship to each other and to the radius of the turn.

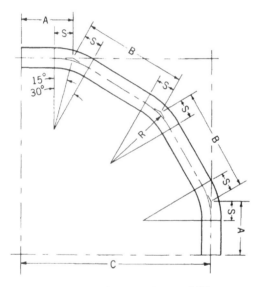

A = Radius of Turn (C) × .268
B = 2A = Radius of Turn (C) × .536

Since the setback for 30° bends is equal to the Radius of Bend × .268, it follows that the Tables for Setback and Length of 30° Bends may be used to find the value of A and B, using as radius, in this case, the Radius of Turn.

EXAMPLE
Find the length of pipe required to make a 90° turn of 24-inch radius with three 30° bends of 11-inch radius.

The 90° Turn
One Pipe, Three-30° Bends
(Continued)

SOLUTION

(1) From Tables for Setback and Length of 30° Bends, find for bends of 11-inch radius:

Setback $= 2\frac{15}{16}$ inches

Length of Bend $= 5\frac{3}{4}$ inches

(2) Using the Setback column in the same table, find for radius of turn equal to 24 inches:

$A = 6\frac{7}{16}$ inches

$B = 2A = 12\frac{7}{8}$ inches

(3) From a point $6\frac{7}{16}$ inches from end of pipe, measure back $2\frac{15}{16}$ inches. This marks beginning of first bend.

(4) From this point, measure forward $5\frac{3}{4}$ inches and mark end of first bend.

(5) Length of straight pipe between bends is equal to length B minus twice the setback. Therefore, from end of first bend measure forward:

$$12\frac{7}{8} - (2 \times 2\frac{15}{16}) = 7 \text{ inches}$$

This marks the beginning of the second bend.

(6) From this point, measure forward $5\frac{3}{4}$ inches and mark end of second bend.

(7) Again, length of straight pipe between bends is length B minus twice the setback. Therefore, measure forward 7 inches and mark beginning of third bend.

(8) Measure forward $5\frac{3}{4}$ inches and mark end of third bend.

(9) Length of straight pipe at the end will be length A minus the setback, or $6\frac{7}{16} - 2\frac{15}{16} = 3\frac{1}{2}$ inches. Measured forward from end of third bend, this marks end of pipe.

(10) Total length of pipe required, therefore, is:

$$6\frac{7}{16} - 2\frac{15}{16} + 5\frac{3}{4} + 7 + 5\frac{3}{4} + 7 + 5\frac{3}{4} + 3\frac{1}{2}$$
$$= 38\frac{1}{4} \text{ inches}$$

Bending Pipe Around a Circle

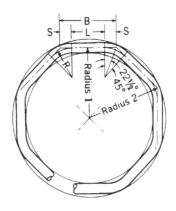

Bending radius, R, may be any convenient length shorter than Radius 1

Length (L) of Straight Pipe between Bends
= B — (2 × Setback)

No. of Sides	Angle of Pipe Bends, Degrees	To Find Distance (B) between Centerline Intersections	
		For Pipe Outside Circle Multiply Rad. 1 by	For Pipe Within Circle Multiply Rad. 2 by
4	90	2.0	1.414
5	72	1.453	1.175
6	60	1.155	1.000
8	45	.828	.765
9	40	.728	.685
10	36	.650	.618
12	30	.536	.518
16	22½	.398	.390
20	18	.317	.313
24	15	.263	.261

Bending Pipe Around a Circle
(Continued)

EXAMPLE

What is the length of 2-inch pipe required to bend one complete octagonal (8-sided) loop outside a circle of 48-inch diameter? Let radius of bend, R, be 10 inches.

SOLUTION

(1) For 8-sided loop:
Angle of pipe bends $= 45°$
Total length of pipe for complete loop $=$
8 $\times$ length of straight pipe between bends $+$
8 $\times$ length of 45° bend of radius R

(2) Radius 1 $=$ ½ diameter of circle $+$ ½ outside diameter of pipe. For 2-inch pipe, outside diameter $=$ 2⅜ inches. Therefore
Radius 1 $=$ ½ $\times$ 48 $+$ ½ $\times$ 2⅜ $=$ 25³⁄₁₆ inches

(3) For 45° bends, distance B $=$ Radius 1 $\times$ .828 $=$ 25³⁄₁₆ $\times$ .828 $=$ 20¹³⁄₁₆ inches.

(4) From Tables for Setback and Length of 45° Bends, for radius of bend (R) equal to 10 inches:
Setback $=$ 4⅛ inches
Length of bend $=$ 7¹³⁄₁₆ inches

(5) Length of straight pipe between bends is equal to distance B — (2 $\times$ setback), or 20¹³⁄₁₆ — (2 $\times$ 4⅛) $=$ 12⁹⁄₁₆ inches.

(6) Therefore, total length of 2-inch pipe required for one complete octagonal loop $=$ 8 $\times$ 12⁹⁄₁₆ $+$ 8 $\times$ 7¹³⁄₁₆ $=$ 163 inches.

Bending Pipe Around a Circle
(Continued)

EXAMPLE

What length of 2-inch pipe would be required to bend one complete pentagonal (5-sided) loop within a circle of 36-inch diameter? Let radius of bend, R, be 8 inches.

SOLUTION

(1) For 5-sided loop:

Angle of pipe bends $= 72°$

Total length of pipe for one loop $=$

5 $\times$ length of straight pipe between bends $+$

5 $\times$ length of 72° bend of radius R

Note: In actual practice, Radius 2 may be taken to the centerline of bend, instead of to the intersection of straight-leg centerlines. This permits simpler calculations, and is an assumption on the safe side when working within any given diameter.

(2) Radius 2 $=$ ½ diameter of circle $—$ ½ outside diameter of pipe. For 2-inch pipe, outside diameter $=$ 2⅜ inches. Therefore

Radius 2 $=$ ½ $\times$ 36 $—$ ½ $\times$ 2⅜ $=$ 16¹³⁄₁₆ inches

(3) For 72° bends, distance B $=$ Radius 2 $\times$ 1.175 $=$ 16¹³⁄₁₆ $\times$ 1.175 $=$ 19¾ inches.

(4) From Tables for Setback and Length of 72° Bends, for radius of bend (R) equal to 8 inches:

Setback $=$ 5¹³⁄₁₆ inches

Length of bend $=$ 10¹⁄₁₆ inches

(5) Length of straight pipe between bends is equal to distance B $—$ (2 $\times$ setback), or 19¾ $—$ (2 $\times$ 5¹³⁄₁₆) $=$ 8⅛ inches.

(6) Therefore, total length of 2-inch pipe required for one complete pentagonal loop $=$

5 $\times$ 8⅛ $+$ 5 $\times$ 10¹⁄₁₆ $=$ 90¹⁵⁄₁₆ inches

Minimum Bending Radius for Standard Weight Pipe

Pipe Size, Inches	Steel Pipe	Wrought Iron Pipe
	Shortest Radius to Which Pipe Can be Bent, Inches‡	
¼	1	—
⅜	1¼	—
½	1½*	1⅜†
¾	1¾*	1¾†
1	2*	2⅛†
1¼	2¼*	2¾†
1½	2½*	3½†
2	3*	5½†
2½	5	10
3	8	12
3½	10	14
4	12	16
5	18	20
6	22	26
8	30	30
10	36	36
12	46	46

‡ Table applies to cold bending and where some degree of flattening at bend is allowed.

* Values from National Tube Co. This company recommends a minimum advisable bend radius five times the nominal pipe size.

† Values from A. M. Byers Co. This company recommends a minimum advisable bend radius five times the nominal pipe size for sizes from 2½ to 4 in., incl. Over 2½ in. wrought iron should be bent hot for smaller radii.

Wrinkle Bends

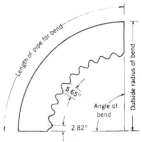

Wrinkle bends can be made with steel or copper pipe by heating one or more narrow bands of the pipe at right angles to the pipe and extending a little more than halfway around the pipe, then pulling the free end of the pipe to form wrinkles where the pipe was heated.

To determine the length of pipe required for a wrinkle bend use the formula

Length =

$$\frac{\text{Outside Radius of Bend} \times \text{Degrees of Angle of Bend}}{57.3}$$

The number of wrinkles is determined by dividing the degrees of bend by the number of degrees per wrinkle.

For example, if a 90 degree bend is to be made with an outside radius of 36 inches, and there is to to be a wrinkle for each 9 degrees, then

Length = (36 × 90) ÷ 57.3 = 56.54 inches

and the number of wrinkles is 90 ÷ 9 = 10.

In this case, then the 56.54 inch length of pipe would be 56.54 ÷ 10 = 5.65 between wrinkles. Allow half this distance at each end of the bend so that there would be 4 spaces of 5.65″ and 2 of 2.82″ each, one at each end.

Common Offset Connections

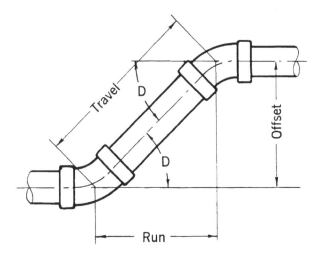

Formulas for calculating offset connections are given on the next page.

Tables of calculated values of travel and run for common offset angles are given on pages following.

Fitting allowances and laying-in dimensions are given in the section on piping dimensions.

EXAMPLE

What would be the length of $2\frac{1}{2}$-inch cast iron pipe required for travel between two 45° elbows so that the offset would be 6 inches?

SOLUTION

(1) Find in table of calculated travel and run, center to center of 45° elbows (or calculate by formula): for 6-inch offset, travel $= 8\frac{1}{2}$ inches

(2) Find in table of laying lengths of 45° elbows (in section on dimensions of screwed fittings) that $2\frac{1}{2}$-inch pipe will extend to within 1 inch of fitting center

(3) From $8\frac{1}{2}$-inch travel, subtract 1 inch at each end to find length of pipe $= 6\frac{1}{2}$ inches.

Common Offset Connections
Two or More Pipes, Equal Spread

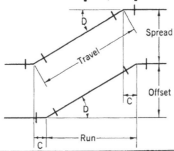

Angle D	To Find	Multiply	by
22½°	Offset = Run	×	.414
	Run = Offset	×	2.414
	Travel = Offset	×	2.613
	C = Spread	×	.199
30°	Offset = Run	×	.577
	Run = Offset	×	1.732
	Travel = Offset	×	2.000
	C = Spread	×	.268
45°	Offset = Run	×	1.000
	Run = Offset	×	1.000
	Travel = Offset	×	1.414
	C = Spread	×	.414
60°	Offset = Run	×	1.732
	Run = Offset	×	.577
	Travel = Offset	×	1.155
	C = Spread	×	.577
Any Angle (See Trig. Tables)	Offset = Run	×	Tan D
	Run = Offset	×	Cot D
	Travel = Offset	×	Csc D
	Offset = Travel	×	Sin D
	C = Spread	×	Tan ½ D

Travel and Run
Center to Center of
11¼° Elbows

Offset, Inches	Travel, Inches	Run, Inches
¼	1⁵⁄₁₆	1¼
½	2⁹⁄₁₆	2½
¾	3⅞	3¾
1	5⅛	5
2	10¼	10¹⁄₁₆
3	15⅜	15¹⁄₁₆
4	20½	20⅛
5	25⅝	25⅛
6	30¾	30³⁄₁₆
7	35⅞	35³⁄₁₆
8	41	40³⁄₁₆
9	46¹⁄₁₆	45¼
10	51³⁄₁₆	50¼
11	56⁵⁄₁₆	55⁵⁄₁₆
12	61⁷⁄₁₆	60⁵⁄₁₆
13	66⁹⁄₁₆	65⅜
14	71¹¹⁄₁₆	70⅜
15	76¹³⁄₁₆	75⁷⁄₁₆

$$\text{Offset} = \text{Run} \times .199$$
$$\text{Run} = \text{Offset} \times 5.027$$
$$\text{Travel} = \text{Offset} \times 5.126$$
$$\text{C} = \text{Spread} \times .099$$

Travel and Run, Center to Center of 22½° Elbows

Offset, Inches	Travel, Inches	Run, Inches
* ¼	⅝	⅝
½	1⁵⁄₁₆	1³⁄₁₆
¾	1¹⁵⁄₁₆	1¹³⁄₁₆
1	2⅝	2⁷⁄₁₆
2	5¼	4¹³⁄₁₆
3	7⅞	7¼
4	10⁷⁄₁₆	9⅝
5	13¹⁄₁₆	12¹⁄₁₆
6	15¹¹⁄₁₆	14½
7	18⅝	16⅞
8	20⅞	19⁵⁄₁₆
9	23½	21¹¹⁄₁₆
10	26⅛	24⅛
11	28¾	26⁹⁄₁₆
12	31⅜	28¹⁵⁄₁₆
13	33¹⁵⁄₁₆	31⅜
14	36⁹⁄₁₆	33¹³⁄₁₆
15	39³⁄₁₆	36³⁄₁₆

* For values not shown, simply add components.

For example, Travel for 15¼-inch offset would be: 31⅜ + 7⅞ + ⅝ = 39⅞ inches (adding values for 12-, 3-, and ¼-inch offsets).

Travel and Run,
Center to Center of
30° Elbows

Offset, Inches	Travel, Inches	Run, Inches
* $\frac{1}{4}$	$\frac{1}{2}$	$\frac{7}{16}$
$\frac{1}{2}$	1	$\frac{7}{8}$
$\frac{3}{4}$	$1\frac{1}{2}$	$1\frac{5}{16}$
1	2	$1\frac{3}{4}$
2	4	$3\frac{7}{16}$
3	6	$5\frac{3}{16}$
4	8	$6\frac{15}{16}$
5	10	$8\frac{5}{8}$
6	12	$10\frac{3}{8}$
7	14	$12\frac{1}{8}$
8	16	$13\frac{7}{8}$
9	18	$15\frac{9}{16}$
10	20	$17\frac{5}{16}$
11	22	19
12	24	$20\frac{13}{16}$
13	26	$22\frac{1}{2}$
14	28	$24\frac{1}{4}$
15	30	26

* For values not shown, simply add components.

For example, Run for $23\frac{1}{2}$-inch offset would be: $20\frac{13}{16} + 19 + \frac{7}{8} = 40\frac{11}{16}$ inches (adding values for 12-, 11-, and $\frac{1}{2}$-inch offsets).

Travel and Run, Center to Center of 45° Elbows

Offset or Run, Inches	Travel, Inches	Offset or Run, Inches	Travel, Inches
* ¼	⅜	14	$19\frac{13}{16}$
½	¾	15	$21\frac{1}{4}$
¾	$1\frac{1}{16}$	16	$22\frac{5}{8}$
1	$1\frac{7}{16}$	17	$24\frac{1}{16}$
2	$2\frac{13}{16}$	18	$25\frac{1}{2}$
3	$4\frac{1}{4}$	19	$26\frac{7}{8}$
4	$5\frac{5}{8}$	20	$28\frac{1}{4}$
5	$7\frac{1}{16}$	21	$29\frac{11}{16}$
6	$8\frac{1}{2}$	22	$31\frac{1}{8}$
7	$9\frac{7}{8}$	23	$32\frac{1}{2}$
8	$11\frac{5}{16}$	24	$33\frac{15}{16}$
9	$12\frac{3}{4}$	25	$35\frac{3}{8}$
10	$14\frac{1}{8}$	26	$36\frac{3}{4}$
11	$15\frac{9}{16}$	27	$38\frac{3}{16}$
12	17	36	$50\frac{15}{16}$
13	$18\frac{7}{16}$	48	$67\frac{7}{8}$

* As shown in the sketch at the beginning of this section, the run of a 45° offset is equal to the offset. Small values are shown in the table so that fractional lengths can be figured by simple addition. For example, Travel for 23¼-inch offset would be (by adding travels for 18-, 5-, and ¼-inch offsets):

$$25\frac{1}{2} + 7\frac{1}{16} + \frac{3}{8} = 32\frac{15}{16} \text{ inches}$$

Travel and Run, Center to Center of 60° Elbows

Offset, Inches	Travel, Inches	Run, Inches
* $\frac{1}{4}$	$\frac{5}{16}$	$\frac{1}{8}$
$\frac{1}{2}$	$\frac{9}{16}$	$\frac{5}{16}$
$\frac{3}{4}$	$\frac{7}{8}$	$\frac{7}{16}$
1	$1\frac{1}{8}$	$\frac{9}{16}$
2	$2\frac{5}{16}$	$1\frac{1}{8}$
3	$3\frac{7}{16}$	$1\frac{3}{4}$
4	$4\frac{5}{8}$	$2\frac{5}{16}$
5	$5\frac{3}{4}$	$2\frac{7}{8}$
6	$6\frac{15}{16}$	$3\frac{7}{16}$
7	$8\frac{1}{16}$	$4\frac{1}{16}$
8	$9\frac{1}{4}$	$4\frac{5}{8}$
9	$10\frac{3}{8}$	$5\frac{3}{16}$
10	$11\frac{9}{16}$	$5\frac{3}{4}$
11	$12\frac{11}{16}$	$6\frac{3}{8}$
12	$13\frac{7}{8}$	$6\frac{15}{16}$
13	15	$7\frac{1}{2}$
14	$16\frac{3}{16}$	$8\frac{1}{16}$
15	$17\frac{5}{16}$	$8\frac{5}{8}$

* For values not shown, simply add components.

For example, Travel for $16\frac{3}{4}$-inch offset would be: $13\frac{7}{8} + 4\frac{5}{8} + \frac{7}{8} = 19\frac{3}{8}$ inches (adding travel values for 12-, 4-, and $\frac{3}{4}$-inch offsets).

Travel and Run
Center to Center of
72° Elbows

Offset, Inches	Travel, Inches	Run, Inches
¼	¼	$\frac{1}{16}$
½	½	⅛
¾	$13\frac{}{16}$	¼
1	$1\frac{1}{16}$	$\frac{5}{16}$
2	$2\frac{1}{8}$	⅝
3	$3\frac{1}{8}$	1
4	$4\frac{3}{16}$	$1\frac{5}{16}$
5	$5\frac{1}{4}$	$1\frac{5}{8}$
6	$6\frac{5}{16}$	2
7	$7\frac{3}{8}$	$2\frac{1}{4}$
8	$8\frac{3}{8}$	$2\frac{5}{8}$
9	$9\frac{7}{16}$	$2\frac{15}{16}$
10	$10\frac{1}{2}$	$3\frac{1}{4}$
11	$11\frac{9}{16}$	$3\frac{9}{16}$
12	$12\frac{5}{8}$	$3\frac{7}{8}$
13	$13\frac{5}{8}$	$4\frac{1}{4}$
14	$14\frac{11}{16}$	$4\frac{9}{16}$
15	$15\frac{3}{4}$	$4\frac{7}{8}$

Offset = Run $\times$ 3.078
Run   = Offset $\times$ .325
Travel = Offset $\times$ 1.051
C   = Spread $\times$ .727

Rolling Offset

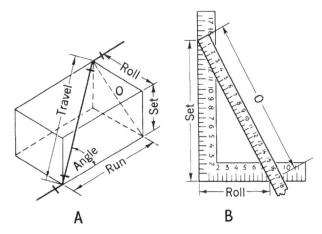

A   B

In the rolling offset, as shown in drawing A above, the plane of the travel is such that the diagonal O is the true offset. There are two methods of finding the travel and run.

(1) Simplified method: Mark off the roll and set on legs of a steel square or any square object as shown in drawing B, above.

Then the straight line distance between these points is the dimension O which is to be read from the tables on preceding pages.

(2) Mathematical method: It can be shown by trigonometry, that

$$O^2 = (Roll)^2 + (Set)^2 \text{ and } O = \sqrt{(Roll)^2 + (Set)^2}$$

Tables of square roots of numbers are given elsewhere for easy solution of this formula.

The same formula can be applied to any offset problem by changing the terms to read:

$$Travel = \sqrt{(Offset)^2 + (Run)^2}$$

Hence, in the rolling offset above,

$$Travel = \sqrt{O^2 + (Run)^2}$$

Two-Pipe 90° Turn
Equal Spread

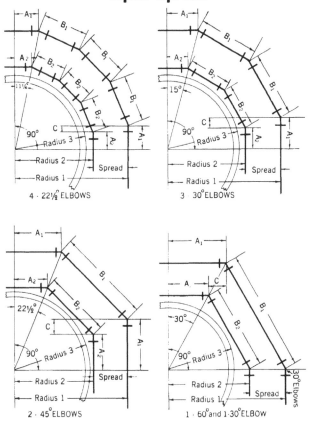

4 · 22½° ELBOWS

3 · 30° ELBOWS

2 · 45° ELBOWS

1 · 60° and 1·30° ELBOW

To Find	Multiply	For 22½° Elbows, by	For 30° Elbows, by	For 45° Elbows, by	For 1-60° & 1-30° El, by
A_1	Radius 1	.199	.268	.414	.577
A_2	Radius 2	.199	.268	.414	.577
B_1	Radius 1	.398	.536	.828	1.155
B_2	Radius 2	.398	.536	.828	1.155
C	Spread	.199	.268	.414	.577

Screwed Pipe Coils
Inside or Outside of Round Tanks

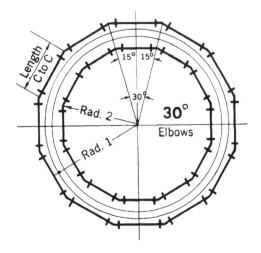

Angle of Fitting Used	No. of Els or Straight Pieces Per Coil	To Find Length of Straight Side, Center to Center of Els	
		In Outside Coil, Multiply Rad. 1 by	In Inside Coil, Multiply Rad. 2 by
22½°	16	.398	.390
30°	12	.536	.518
45°	8	.828	.765
60°	6	1.155	1.000
90°	4	2.000	1.414
A*	$\dfrac{360}{A}$	$2 \times \text{Tan } \frac{1}{2} A$	$2 \times \text{Sin } \frac{1}{2} A$

* A can be any angle by which 360 can be divided evenly. See trigonometric tables for tangents and sines.

Mitering Pipe

A miter cut is made at $\frac{1}{2}$ the angle of the completed joint.

In fabricating welded offset connections the cut is positioned as shown.

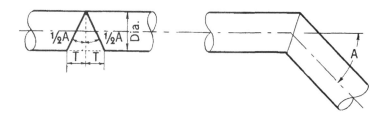

To Make a Mitered Offset at Angle A Equal to	Mark Off Distance T Equal to Dia.* Times	If Nominal Pipe Dia. Is			
		2″	3″	4″	6″
		Distance T in Inches Is			
22½°	.199	.472 ($\frac{1}{2}$)	.696 ($\frac{11}{16}$)	.895 ($\frac{7}{8}$)	1.32 ($1\frac{5}{16}$)
30°	.268	.635 ($\frac{5}{8}$)	.937 ($\frac{15}{16}$)	1.208 ($1\frac{3}{16}$)	1.77 ($1\frac{3}{4}$)
45°	.414	.982 (1)	1.45 ($1\frac{7}{16}$)	1.86 ($1\frac{7}{8}$)	2.76 ($2\frac{3}{4}$)
60°	.577	1.37 ($1\frac{3}{8}$)	2.02 (2)	2.60 ($2\frac{5}{8}$)	3.82 ($3\frac{13}{16}$)
90°	1.00	2.37 ($2\frac{3}{8}$)	3.500 ($3\frac{1}{2}$)	4.500 ($4\frac{1}{2}$)	6.63 ($6\frac{5}{8}$)
Any Angle	Tan $\frac{1}{2}$ A	(Values in parentheses above, are to nearest $\frac{1}{16}$ inch)			

* For greatest accuracy, use actual O.D. of pipe as given in tables in section on pipe standards.

Tangents of all angles are given in trigonometry section.

The 45° Welded Turn
3-Piece

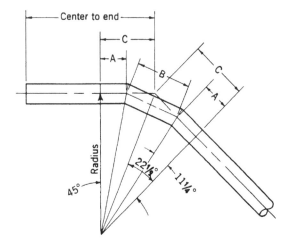

There are 2 mitered joints in a 3-piece 45° turn. Each joint is 45/2 = 22½°, and each piece is mitered at 22½/2 = 11¼°.

In the drawing:

A = Radius × .1989
B = 2A = Radius × .3978
C = Radius × .4142

Calculated values for radius of turn from 1 to 48 inches are given in tables beginning next page.

The mitered pipe is always cut at 11¼° to make two 22½° joints in a 3-piece 45° turn. The angle of the miter cut always has its apex at the center from which the radius of the turn is drawn.

If the 3-piece 45° turn is followed by a reversed 3-piece 45° turn to form a 45° mitered offset, values of travel and run for the required offset distance can be found using the procedure and table for 45° offset bends found in the section on Pipe Bending.

The 45° Welded Turn
3-Piece

Radius of Turn, Inches	Length A, Inches	Length B, Inches	Length C, Inches
1	$\frac{3}{16}$	$\frac{3}{8}$	$\frac{7}{16}$
2	$\frac{3}{8}$	$\frac{13}{16}$	$\frac{13}{16}$
3	$\frac{5}{8}$	$1\frac{3}{16}$	$1\frac{1}{4}$
4	$\frac{13}{16}$	$1\frac{5}{8}$	$1\frac{5}{8}$
5	1	2	$2\frac{1}{16}$
6	$1\frac{3}{16}$	$2\frac{3}{8}$	$2\frac{7}{16}$
7	$1\frac{3}{8}$	$2\frac{3}{4}$	$2\frac{7}{8}$
8	$1\frac{5}{8}$	$3\frac{3}{16}$	$3\frac{1}{4}$
9	$1\frac{13}{16}$	$3\frac{9}{16}$	$3\frac{3}{4}$
10	2	4	$4\frac{1}{8}$
11	$2\frac{3}{16}$	$4\frac{3}{8}$	$4\frac{1}{2}$
12	$2\frac{3}{8}$	$4\frac{3}{4}$	$4\frac{15}{16}$
13	$2\frac{9}{16}$	$5\frac{3}{16}$	$5\frac{5}{16}$
14	$2\frac{3}{4}$	$5\frac{9}{16}$	$5\frac{3}{4}$
15	3	6	$6\frac{1}{8}$
16	$3\frac{3}{16}$	$6\frac{3}{8}$	$6\frac{9}{16}$
17	$3\frac{3}{8}$	$6\frac{3}{4}$	7
18	$3\frac{9}{16}$	$7\frac{3}{16}$	$7\frac{7}{16}$
19	$3\frac{3}{4}$	$7\frac{9}{16}$	$7\frac{7}{8}$
20	4	$7\frac{15}{16}$	$8\frac{3}{16}$
21	$4\frac{3}{16}$	$8\frac{3}{8}$	$8\frac{5}{8}$
22	$4\frac{3}{8}$	$8\frac{3}{4}$	$9\frac{1}{8}$
23	$4\frac{9}{16}$	$9\frac{1}{8}$	$9\frac{7}{16}$
24	$4\frac{3}{4}$	$9\frac{9}{16}$	$9\frac{11}{16}$

The 45° Welded Turn
3-Piece

Radius of Turn, Inches	Length A, Inches	Length B, Inches	Length C, Inches
25	5	$9\frac{15}{16}$	$10\frac{1}{4}$
26	$5\frac{3}{16}$	$10\frac{3}{8}$	$10\frac{11}{16}$
27	$5\frac{3}{8}$	$10\frac{3}{4}$	$11\frac{1}{16}$
28	$5\frac{9}{16}$	$11\frac{1}{8}$	$11\frac{1}{2}$
29	$5\frac{3}{4}$	$11\frac{1}{2}$	$11\frac{7}{8}$
30	6	$11\frac{15}{16}$	$12\frac{7}{16}$
31	$6\frac{3}{16}$	$12\frac{5}{16}$	$12\frac{13}{16}$
32	$6\frac{3}{8}$	$12\frac{3}{4}$	$13\frac{1}{4}$
33	$6\frac{9}{16}$	$13\frac{1}{8}$	$13\frac{11}{16}$
34	$6\frac{3}{4}$	$13\frac{1}{2}$	$14\frac{1}{16}$
35	$6\frac{15}{16}$	$13\frac{15}{16}$	$14\frac{1}{2}$
36	$7\frac{3}{16}$	$14\frac{5}{16}$	$14\frac{15}{16}$
37	$7\frac{3}{8}$	$14\frac{11}{16}$	$15\frac{5}{16}$
38	$7\frac{9}{16}$	$15\frac{1}{8}$	$15\frac{3}{4}$
39	$7\frac{3}{4}$	$15\frac{1}{2}$	$16\frac{1}{8}$
40	$7\frac{15}{16}$	$15\frac{15}{16}$	$16\frac{9}{16}$
41	$8\frac{3}{16}$	$16\frac{5}{16}$	17
42	$8\frac{3}{8}$	$16\frac{11}{16}$	$17\frac{3}{8}$
43	$8\frac{9}{16}$	$17\frac{1}{8}$	$17\frac{13}{16}$
44	$8\frac{3}{4}$	$17\frac{1}{2}$	$18\frac{3}{16}$
45	$8\frac{15}{16}$	$17\frac{15}{16}$	$18\frac{5}{8}$
46	$9\frac{3}{16}$	$18\frac{5}{16}$	$19\frac{1}{16}$
47	$9\frac{3}{8}$	$18\frac{11}{16}$	$19\frac{7}{16}$
48	$9\frac{9}{16}$	$19\frac{1}{8}$	$19\frac{7}{8}$

The 60° Welded Turn
3-Piece

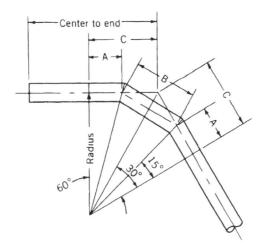

There are 2 mitered joints in a 3-piece 60° turn. Each joint is 60/2 = 30°, and each piece is mitered at 30/2 = 15°.

In the drawing:
    A = Radius × .2679
    B = 2A = Radius × .5358
    C = Radius × .5774

Calculated values for radius of turn from 1 to 48 inches are given in tables beginning next page.

The mitered pipe is always cut at 15° to make two 30° joints in a 3-piece 60° turn. The angle of the miter cut always has its apex at the center from which the radius of the turn is drawn.

If the 3-piece 60° turn is followed by a reversed 3-piece 60° turn to form a 60° mitered offset, values of travel and run for the required offset distance can be found using the procedure and table for 60° offset bends found in the section on Pipe Bending.

The 60° Welded Turn
3-Piece

Radius of Turn, Inches	Length A, Inches	Length B, Inches	Length C, Inches
1	¼	⁹⁄₁₆	⁹⁄₁₆
2	⁹⁄₁₆	1¹⁄₁₆	1⅛
3	¹³⁄₁₆	1⅝	1¾
4	1¹⁄₁₆	2⅛	2⁵⁄₁₆
5	1⁵⁄₁₆	2¹¹⁄₁₆	2⅞
6	1⅝	3³⁄₁₆	3½
7	1⅞	3¾	4¹⁄₁₆
8	2⅛	4⁵⁄₁₆	4⅝
9	2⁷⁄₁₆	4⅞	5³⁄₁₆
10	2¹¹⁄₁₆	5⅜	5¹³⁄₁₆
11	2¹⁵⁄₁₆	5⅞	6⅜
12	3³⁄₁₆	6⁷⁄₁₆	6¹⁵⁄₁₆
13	3½	6¹⁵⁄₁₆	7½
14	3¾	7½	8⅛
15	4	8	8¹¹⁄₁₆
16	4⁵⁄₁₆	8⁹⁄₁₆	9¼
17	4⁹⁄₁₆	9⅛	9⅞
18	4¹³⁄₁₆	9⅝	10⁷⁄₁₆
19	5¹⁄₁₆	10³⁄₁₆	11
20	5⅜	10¾	11⁹⁄₁₆
21	5⅝	11¼	12³⁄₁₆
22	5⅞	11¹³⁄₁₆	12¾
23	6³⁄₁₆	12⁵⁄₁₆	13⁵⁄₁₆
24	6⁷⁄₁₆	12⅞	13¹⁵⁄₁₆

The 60° Welded Turn
3-Piece

Radius of Turn, Inches	Length A, Inches	Length B, Inches	Length C, Inches
25	$6\frac{11}{16}$	$13\frac{3}{8}$	$14\frac{1}{2}$
26	7	$13\frac{15}{16}$	$15\frac{1}{16}$
27	$7\frac{1}{4}$	$14\frac{1}{2}$	$15\frac{5}{8}$
28	$7\frac{1}{2}$	15	$16\frac{1}{4}$
29	$7\frac{3}{4}$	$15\frac{1}{2}$	$16\frac{13}{16}$
30	$8\frac{1}{16}$	$16\frac{1}{16}$	$17\frac{3}{8}$
31	$8\frac{5}{16}$	$16\frac{5}{8}$	$17\frac{7}{8}$
32	$8\frac{9}{16}$	$17\frac{1}{8}$	$18\frac{1}{2}$
33	$8\frac{7}{8}$	$17\frac{11}{16}$	$19\frac{1}{16}$
34	$9\frac{1}{8}$	$18\frac{3}{16}$	$19\frac{11}{16}$
35	$9\frac{3}{8}$	$18\frac{3}{4}$	$20\frac{1}{4}$
36	$9\frac{5}{8}$	$19\frac{5}{16}$	$20\frac{13}{16}$
37	$9\frac{15}{16}$	$19\frac{7}{8}$	$21\frac{3}{8}$
38	$10\frac{3}{16}$	$20\frac{3}{8}$	$21\frac{7}{8}$
39	$10\frac{7}{16}$	$20\frac{15}{16}$	$22\frac{1}{2}$
40	$10\frac{11}{16}$	$21\frac{7}{16}$	$23\frac{1}{8}$
41	11	22	$23\frac{11}{16}$
42	$11\frac{1}{4}$	$22\frac{9}{16}$	$24\frac{1}{4}$
43	$11\frac{1}{2}$	$23\frac{1}{16}$	$24\frac{13}{16}$
44	$11\frac{3}{4}$	$23\frac{5}{8}$	$25\frac{7}{16}$
45	$12\frac{1}{16}$	$24\frac{1}{8}$	26
46	$12\frac{5}{16}$	$24\frac{11}{16}$	$26\frac{9}{16}$
47	$12\frac{9}{16}$	$25\frac{3}{16}$	$27\frac{1}{8}$
48	$12\frac{7}{8}$	$25\frac{3}{4}$	$27\frac{11}{16}$

The 90° Welded Turn, 3-Piece

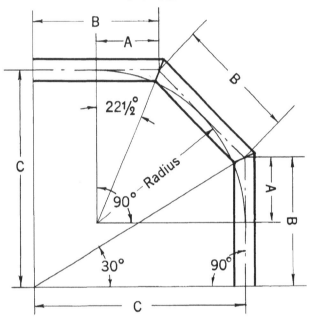

There are 2 mitered joints in a 3-piece 90° turn. Each joint is 90/2 = 45°, and each piece is mitered at 45/2 = 22½°.

In the drawing:

A = Radius × .4142
B = 2A = Radius × .828
C = B × 1.732
B = C × .5774

Notice that, although the drawing shows a 30° angle in the triangle with sides B and C, this angle is used only to determine those sides and does not set the angle of the miter cut. The mitered pipe is always cut at 22½° to make two 45° joints in a 3-piece 90° turn. The angle of the miter cut always has its apex at the center from which the radius of the turn is drawn.

The 90° Welded Turn
3-Piece

Radius of Turn, Inches	Length A, Inches	Length B, Inches
1	$\frac{7}{16}$	$\frac{13}{16}$
2	$\frac{13}{16}$	$1\frac{11}{16}$
3	$1\frac{1}{4}$	$2\frac{1}{2}$
4	$1\frac{11}{16}$	$3\frac{5}{16}$
5	$2\frac{1}{16}$	$4\frac{1}{8}$
6	$2\frac{1}{2}$	5
7	$2\frac{7}{8}$	$5\frac{13}{16}$
8	$3\frac{5}{16}$	$6\frac{5}{8}$
9	$3\frac{3}{4}$	$7\frac{7}{16}$
10	$4\frac{1}{8}$	$8\frac{5}{16}$
11	$4\frac{9}{16}$	$9\frac{1}{8}$
12	5	$9\frac{15}{16}$
13	$5\frac{3}{8}$	$10\frac{3}{4}$
14	$5\frac{13}{16}$	$11\frac{5}{8}$
15	$6\frac{3}{16}$	$12\frac{7}{16}$
16	$6\frac{5}{8}$	$13\frac{1}{4}$
17	$7\frac{1}{16}$	$14\frac{1}{16}$
18	$7\frac{7}{16}$	$14\frac{15}{16}$
19	$7\frac{7}{8}$	$15\frac{3}{4}$
20	$8\frac{5}{16}$	$16\frac{9}{16}$
21	$8\frac{11}{16}$	$17\frac{3}{8}$
22	$9\frac{1}{8}$	$18\frac{1}{4}$
23	$9\frac{1}{2}$	$19\frac{1}{16}$
24	$9\frac{15}{16}$	$19\frac{7}{8}$

The 90° Welded Turn
3-Piece

Radius of Turn, Inches	Length A, Inches	Length B, Inches
25	$10\frac{3}{8}$	$20\frac{11}{16}$
26	$10\frac{3}{4}$	$21\frac{9}{16}$
27	$11\frac{3}{16}$	$22\frac{3}{8}$
28	$11\frac{5}{8}$	$23\frac{3}{16}$
29	12	24
30	$12\frac{7}{16}$	$24\frac{7}{8}$
31	$12\frac{13}{16}$	$25\frac{11}{16}$
32	$13\frac{1}{4}$	$26\frac{1}{2}$
33	$13\frac{11}{16}$	$27\frac{5}{16}$
34	$14\frac{1}{16}$	$28\frac{3}{16}$
35	$14\frac{1}{2}$	29
36	$14\frac{15}{16}$	$29\frac{13}{16}$
37	$15\frac{5}{16}$	$30\frac{5}{8}$
38	$15\frac{3}{4}$	$31\frac{7}{16}$
39	$16\frac{1}{8}$	$32\frac{5}{16}$
40	$16\frac{9}{16}$	$33\frac{1}{8}$
41	$16\frac{15}{16}$	$33\frac{15}{16}$
42	$17\frac{3}{8}$	$34\frac{13}{16}$
43	$17\frac{13}{16}$	$35\frac{5}{8}$
44	$18\frac{3}{16}$	$36\frac{7}{16}$
45	$18\frac{5}{8}$	$37\frac{1}{4}$
46	19	$38\frac{1}{8}$
47	$19\frac{7}{16}$	$38\frac{7}{8}$
48	$19\frac{7}{8}$	$39\frac{3}{4}$

The 90° Welded Turn, 4-Piece

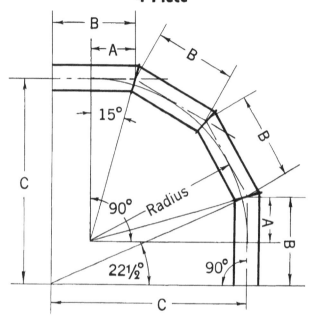

There are three mitered joints in a 4-piece 90° turn. Each joint is 90/3 = 30°, and each miter is 30/2 = 15°.

In the drawing:

A = Radius × .2679
B = 2A = Radius × .536
C = B × 2.4142
B = C × .4142

The drawing shows an angle of 22½° in the triangle with sides B and C. This angle is used only to determine the length of B or C when one or the other side is known. The angle of the miter cut always has its apex at the center from which the radius of the turn is drawn. In a 4-piece, 90° turn, the mitered pipe is always cut at 15° to make three 30° joints.

The 90° Welded Turn
4-Piece

Radius of Turn, Inches	Length A, Inches	Length B, Inches
1	¼	⁹⁄₁₆
2	⁹⁄₁₆	1¹⁄₁₆
3	¹³⁄₁₆	1⅝
4	1¹⁄₁₆	2⅛
5	1⅜	2¹¹⁄₁₆
6	1⅝	3³⁄₁₆
7	1⅞	3¾
8	2⅛	4⁵⁄₁₆
9	2⁷⁄₁₆	4⅞
10	2¹¹⁄₁₆	5⅜
11	2¹⁵⁄₁₆	5⅞
12	3¼	6⁷⁄₁₆
13	3½	6¹⁵⁄₁₆
14	3¾	7½
15	4	8
16	4⁵⁄₁₆	8⁹⁄₁₆
17	4⁹⁄₁₆	9⅛
18	4¹³⁄₁₆	9⅝
19	5⅛	10³⁄₁₆
20	5⅜	10¾
21	5⅝	11¼
22	5⅞	11¹³⁄₁₆
23	6³⁄₁₆	12⁵⁄₁₆
24	6⁷⁄₁₆	12⅞

The 90° Welded Turn
4-Piece

Radius of Turn, Inches	Length A, Inches	Length B, Inches
25	$6^{11}\!/_{16}$	$13\frac{3}{8}$
26	7	$13^{15}\!/_{16}$
27	$7\frac{1}{4}$	$14\frac{1}{2}$
28	$7\frac{1}{2}$	15
29	$7\frac{3}{4}$	$15\frac{1}{2}$
30	$8\frac{1}{16}$	$16\frac{1}{16}$
31	$8^{5}\!/_{16}$	$16\frac{5}{8}$
32	$8^{9}\!/_{16}$	$17\frac{1}{8}$
33	$8\frac{7}{8}$	$17^{11}\!/_{16}$
34	$9\frac{1}{8}$	$18^{3}\!/_{16}$
35	$9\frac{3}{8}$	$18\frac{3}{4}$
36	$9^{11}\!/_{16}$	$19^{5}\!/_{16}$
37	$9^{15}\!/_{16}$	$19^{13}\!/_{16}$
38	$10^{3}\!/_{16}$	$20\frac{3}{8}$
39	$10^{7}\!/_{16}$	$20^{15}\!/_{16}$
40	$10\frac{3}{4}$	$21^{7}\!/_{16}$
41	11	22
42	$11\frac{1}{4}$	$22\frac{1}{2}$
43	$11\frac{1}{2}$	$23\frac{1}{16}$
44	$11^{13}\!/_{16}$	$23\frac{5}{8}$
45	$12\frac{1}{16}$	$24\frac{1}{8}$
46	$12^{5}\!/_{16}$	$24^{11}\!/_{16}$
47	$12\frac{5}{8}$	$25^{3}\!/_{16}$
48	$12\frac{7}{8}$	$25\frac{3}{4}$

Mitering Pipe Around A Circle

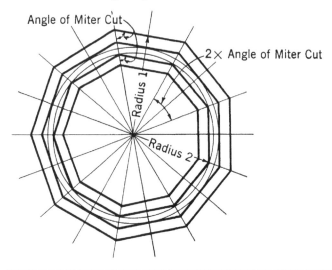

No. of Sides	Angle of Miter Cut, Degrees	To Find Length of Outside Edge of Straight Side	
		Outside Circle Multiply Rad. 1 by	Inside Circle Multiply Rad. 2 by
4	45	2.000	1.414
5	36	1.453	1.175
6	30	1.155	1.000
8	22½	.828	.765
9	20	.728	.685
10	18	.650	.618
12	15	.536	.518
16	11¼	.398	.390
20	9	.317	.313
X	$\dfrac{360}{2X}$	$2 \times \operatorname{Tan}\dfrac{360}{2X}$	$2 \times \operatorname{Sin}\dfrac{360}{2X}$

90° Brackets
Diagonal Leg at 22½°

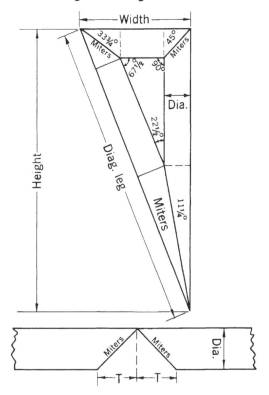

Any material , welded pipe or wood, can be used to construct brackets.

To make pieces, cut as shown:

Two 45°   Miters with T = Dia. × 1.00
Two 33¾° Miters with T = Dia. × 1.50
Two 11¼° Miters with T = Dia. × 5.03
         Width = Height × .4142
        Height = Width × 2.4142
     Diag. Leg = Width × 2.6131

Be sure that T dimensions of each piece are measured on a single centerline.

Dimensions of 90° Brackets
Diagonal Leg at 22½°

Width, Inches	Height, Inches	Diag. Leg, Inches
4	$9^{11}/_{16}$	$10^{7}/_{16}$
6	$14^{1}/_{2}$	$15^{11}/_{16}$
8	$19^{5}/_{16}$	$20^{7}/_{8}$
10	$24^{1}/_{8}$	$26^{1}/_{8}$
12	25	$31^{3}/_{8}$
14	$33^{13}/_{16}$	$36^{9}/_{16}$
16	$38^{5}/_{8}$	$41^{13}/_{16}$
18	$43^{7}/_{16}$	47
20	$48^{5}/_{16}$	$52^{1}/_{4}$
22	$53^{1}/_{8}$	$57^{1}/_{2}$
24	58	$62^{11}/_{16}$
26	$62^{9}/_{16}$	$67^{15}/_{16}$
28	$67^{5}/_{8}$	$73^{3}/_{16}$
30	$72^{7}/_{16}$	$78^{3}/_{8}$
32	$77^{1}/_{4}$	$83^{5}/_{8}$
34	$82^{1}/_{16}$	$88^{7}/_{8}$
36	$86^{15}/_{16}$	$94^{1}/_{16}$
38	$91^{3}/_{4}$	$99^{5}/_{16}$
40	$96^{9}/_{16}$	$104^{1}/_{2}$
42	$101^{3}/_{8}$	$109^{3}/_{4}$
44	$106^{1}/_{4}$	115
46	$111^{1}/_{16}$	$120^{3}/_{16}$
48	$115^{7}/_{8}$	$125^{7}/_{16}$

90° Brackets
Diagonal Leg at 30° or 60°

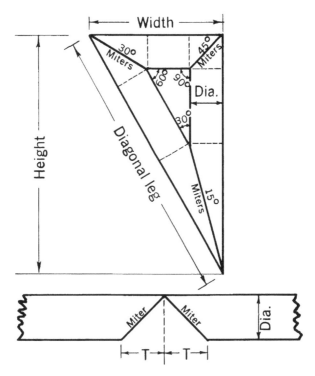

To make pieces, cut as shown:

Two 45° Miters with T = Dia. × 1.00

Two 30° Miters with T = Dia. × 1.73

Two 15° Miters with T = Dia. × 3.73

For bracket with diagonal leg at 30°:

Width = Height × .58

Height = Width × 1.73

Diag. Leg = Width × 2.00

Note: Setting short leg vertical against wall inclines diagonal leg at 60° and exchanges width and height terms in above formulas and in table on opposite page.

Dimensions of 90° Brackets
Diagonal Leg at 30°

Width, Inches	Height, Inches	Diag. Leg, Inches
4	$6^{15}/_{16}$	8
6	$10^3/_8$	12
8	$13^7/_8$	16
10	$17^5/_{16}$	20
12	$20^{13}/_{16}$	24
14	$24^1/_4$	28
16	$27^{11}/_{16}$	32
18	$31^3/_{16}$	36
20	$34^5/_8$	40
22	$38^1/_8$	44
24	$41^9/_{16}$	48
26	45	52
28	$48^7/_{16}$	56
30	52	60
32	$55^7/_{16}$	64
34	$58^7/_8$	68
36	$62^3/_8$	72
38	$65^{13}/_{16}$	76
40	$69^1/_4$	80
42	$72^3/_4$	84
44	$76^3/_{16}$	88
46	$79^{11}/_{16}$	92
48	$83^1/_8$	96

90° Brackets
Diagonal Leg at 45°

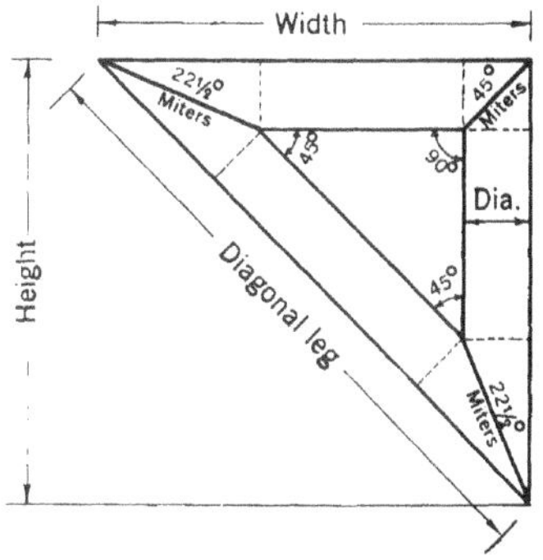

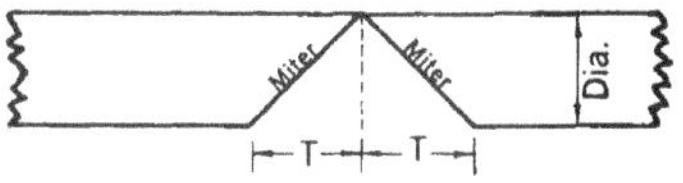

To make pieces, cut as shown:

Two 45° Miters with T = Dia. × 1.00

Four 22½° Miters with T = Dia. × 2.4142

Width = Height

Diag. Leg = Width × 1.4142

Dimensions of Pieces, Inches					
Width or Height	Length of Diag. Leg	Width or Height	Length of Diag. Leg	Width or Height	Length of Diag. Leg
4	5⅝	16	22⅝	28	39 9/16
6	8½	18	25 7/16	30	42 7/16
8	11 5/16	20	28 5/16	32	45¼
10	14⅛	22	31⅛	34	48 1/16
12	17	24	33 15/16	36	50 15/16
14	19 13/16	26	36¾		

Construction of 90° Brackets
Diagonal Leg at Any Angle

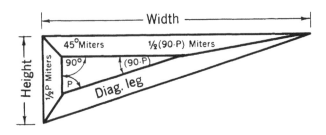

Pitch of diagonal leg is at angle P from wall.

Width = Height × Tan P = Diag. Leg × Sin P

Height = Width × Cot P = Diag. Leg × Cos P

Diag. Leg = Height × Sec P = Width × Csc P

For Tan, Sin, etc. see tables in Trigonometry section.

Pieces for the bracket are cut as follows:

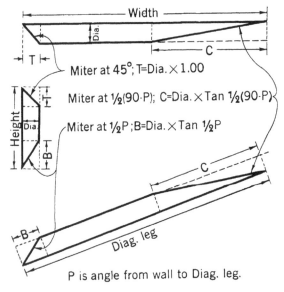

Miter at 45°; T=Dia. × 1.00

Miter at ½(90-P); C=Dia. × Tan ½(90-P)

Miter at ½P; B=Dia. × Tan ½P

P is angle from wall to Diag. leg.

In mitering round pieces, care must be taken that the cuts on both ends are based on a single centerline.

Dividing the Pipe Circumference into Equal Parts

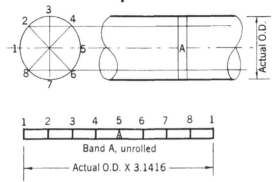

Band A, unrolled

Actual O.D. X 3.1416

Fabrication of a mitered joint may require division of the pipe circumference into equal segments either for marking pipe in the field prior to cutting and welding or for the laying out of templates on paper for later use in marking pipe for cutting.

In the field, simply wrap a band of paper around the pipe and divide the band evenly by folding. The band can then be reapplied to the pipe and the fold marks used as guides in marking the pipe. A piece of angle iron, with true sides, can be laid along the pipe as a guide for making lines along the pipe.

To lay out a template, use is made of a band similar to that shown in the drawing above. On one side of the band, an edge is constructed in a manner shown on subsequent pages. The length of the band is drawn equal to the actual circumference of the pipe. Hence the actual outside rather than nominal pipe diameter is used in calculating:

$$\text{Length of segment} = \frac{\text{Actual O.D.} \times 3.1416}{\text{Number of Segments}}$$

Calculated values are on next page.

Length of Equal Segments of a Pipe Circumference

(Schedule 40 Steel Pipe)

Nominal Pipe Size, Inches	Actual O. D., Inches	Actual Circum., Inches	Number of Segments				
			4	6	8	10	12
			Length of Segment, Inches				
1¼	1.660	5.215	1.30	.87	.65	.52	.43
1½	1.900	5.969	1.49	1.00	.75	.60	.50
2	2.375	7.461	1.87	1.24	.93	.75	.62
2½	2.875	9.032	2.26	1.51	1.13	.90	.75
3	3.500	10.996	2.75	1.67	1.37	1.10	.92
3½	4.000	12.566	3.14	2.10	1.57	1.26	1.05
4	4.500	14.137	3.54	2.36	1.75	1.41	1.18
5	5.563	17.477	4.37	2.92	2.18	1.75	1.46
6	6.625	20.813	5.20	3.47	2.60	2.08	1.73
8	8.625	27.096	6.77	4.52	3.39	2.71	2.26
10	10.750	33.772	8.45	5.60	4.23	3.38	2.82
12	12.750	40.055	10.00	6.67	5.00	4.01	3.34
14	14.000	44.000	11.00	7.35	5.51	4.40	3.66
16	16.000	50.375	12.60	8.40	6.30	5.04	4.19
18	18.000	56.549	14.13	9.41	7.06	5.65	4.70
20	20.000	62.832	15.70	10.48	7.85	6.28	5.22

Decimal equivalents of fractions of an inch are shown on the next page. Lengths of segments in standard fractions of an inch can be found to the nearest $\frac{1}{64}$ inch as follows:

If the segment length is given above as, for example, 3.14 inches, the decimal fraction .14 is found on the next page to be $\frac{9}{64}$ and the segment length is, therefore, $3\frac{9}{64}$ inches. The nearest practical value is seen to be $3\frac{1}{8}$ inches.

Decimal Equivalents of Standard Fractions

1/64	0.015625	33/64	0.515625
1/32	0.03125	17/32	0.53125
3/64	0.046875	35/64	0.546875
1/16	0.0625	9/16	0.5625
5/64	0.078125	37/64	0.578125
3/32	0.09375	19/32	0.59375
7/64	0.109375	39/64	0.609375
1/8	0.125	5/8	0.625
9/64	0.140625	41/64	0.640625
5/32	0.15625	21/32	0.65625
11/64	0.171875	43/64	0.671875
3/16	0.1875	11/16	0.6875
13/64	0.203125	45/64	0.703125
7/32	0.21875	23/32	0.71875
15/64	0.234375	47/64	0.734375
1/4	0.25	3/4	0.75
17/64	0.265625	49/64	0.765625
9/32	0.28125	25/32	0.78125
19/64	0.296875	51/64	0.796875
5/16	0.3125	13/16	0.8125
21/64	0.328125	53/64	0.828125
11/32	0.34375	27/32	0.84375
23/64	0.359375	55/64	0.859375
3/8	0.375	7/8	0.875
25/64	0.390625	57/64	0.890625
13/32	0.40625	29/32	0.90625
27/64	0.421875	59/64	0.921875
7/16	0.4375	15/16	0.9375
29/64	0.453125	61/64	0.953125
15/32	0.46875	31/32	0.96875
31/64	0.484375	63/64	0.984375
1/2	0.5	1	1.

Construction of Miter Templates

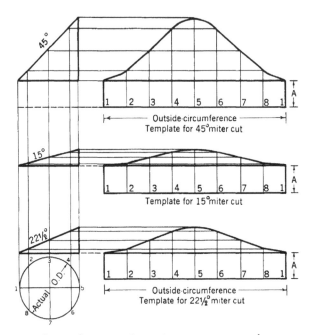

Outside-circumference
Template for 45°miter cut

Template for 15°miter cut

Outside-circumference
Template for 22½°miter cut

Templates for mitering pipe at any angle are constructed as shown above.

(1) Draw circle with diameter equal to O.D. of pipe. (2) Draw horizontal line equal to circumference of circle. (3) Divide circle and horizontal line into equal segments and number divisions as shown. (4) Erect perpendiculars at numbered points. (5) Draw line at desired angle across verticals from circle. (6) Draw horizontals from angle line intersections. (7) Draw template edge through intersections of horizontals and verticals. Dimension, A, can be any convenient length. See previous pages for segment lengths.

In the illustration, 8 segments are used for convenience and clarity. For greater accuracy in drawing the curved edge of the template, a larger number of segments will provide more intersection points.

Construction of Templates for 90° Tee Branch

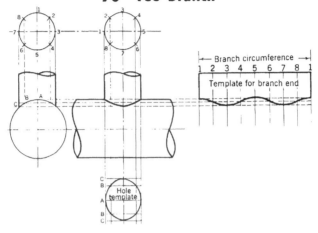

Construction of templates for marking 90° tee welded joints is shown above.

Notice that the hole template can be made by projecting either the inside or outside circumference of the branch onto the curved surface of the header. The choice depends on the welding method to be used. If the outside circumference is to be used, the actual O.D. plus $\frac{1}{8}$ inch should be used as the branch diameter, and the header hole should be beveled at 45°. This is to allow clearance for insertion of welding metal.

Where the branch is smaller than the header, as in the illustration, the curved distance from A to C around the header circumference, which is flattened out in drawing the hole template, can be obtained with sufficient accuracy by taking off chords A-B and B-C with dividers rather than actually measuring the arcs. For large pipe, use more segments.

Actual diameters and circumferences must be used in making templates, which must be drawn full scale, of course.

Construction of Templates for
45° Branch Connection

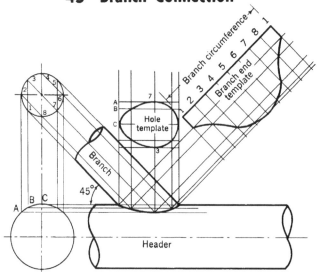

Construction of templates for marking **45°** branch welded joints is clear from the illustration above and description on previous pages. Points to be considered are:

(1) The hole template can be drawn by projecting either the inside or outside circumference of the branch onto the corresponding circumference of the header in the same manner as shown.

(2) Notice that when the hole template is made the dimensions A-B and B-C are segments of the header circumference flattened out so that the distance from 3 to 7 is actually greater than the diameter of the branch. Sufficient accuracy is obtained by using chords A-B and B-C rather than measuring the true length of circumference segments.

(3) The branch end template is always based on the outside circumference.

Field Method for Marking
Welded Branch Connections

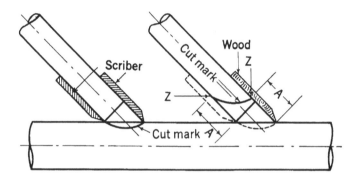

Pipe can be marked in the field for cutting and welding branch connections at any desired angle by the method shown above.

(1) Support the branch (or a piece of pipe of the same size) at the desired angle touching the header. Lay a scriber or a long piece of soapstone sharpened to a flat edge, against the branch and move the stone around the branch keeping the point in contact with the header. This will mark the line for cutting the hole in the header.

(2) To mark the branch, use a piece of wood sharpened to a flat edge. Mark off a distance, A, from the sharpened edge such that a marker held at Z will not go off the end of the pipe as the wooden piece is moved around the branch with the sharp end in contact with the header.

(3) Cutting torch, when cutting the hole in the header, should be held at the same angle as the branch insertion. A radial cut is made on the branch end.

(4) Header hole should be beveled for penetration of welding metal.

Construction of Templates for a True Wye

Drawing of templates for a true wye, with main and branches of equal size and branches at 30° from main centerline is illustrated on the opposite page.

Templates for branch ends are identical. Hence only one is drawn.

Construction procedure is the same regardless of the angle of the wye.

Circumferences can be divided into any number of segments as described on previous pages. Accuracy of the template increases as more segments are taken.

If an uneven number of segments is used, a full circle on the actual O.D. must be used in projecting points for templates. It is usually more convenient to use an even number of segments where the joint is symmetrical.

Arrange centerlines so that high points and low points on the template are clearly defined, as they are at 1, 4, and 7 in the drawing opposite.

Note that curves are smoothed throughout and are tangent to the guide lines at lines 1 and 7. Points are developed at lines numbered 4 in the drawing opposite.

Fabrication of Branch Joints

Intersection joints, such as tees and wyes, are usually the most difficult to weld. In order to be sure the bursting strength of the connection is equal to that of the piping, a factory made fitting is preferred, or one that is shop welded. Unreinforced welded branches are not recommended for use at pressures above 75% of the pipe rating.

The pipe edges to be welded should be beveled so that a V-shaped channel of about 45° occurs for filling and penetration by weld metal.

Construction of Templates for a True Wye

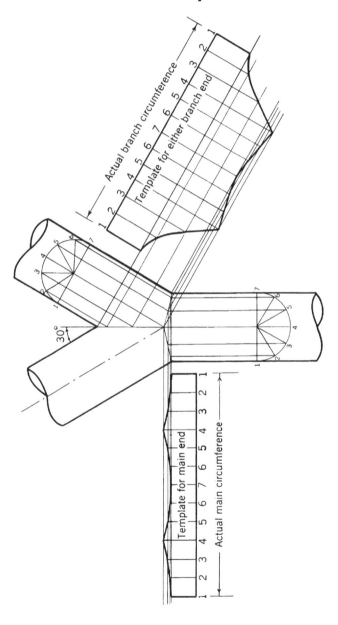

Dimensions of Schedule 40 (Standard Weight) Steel and Wrought Iron Pipe

Nominal Diameter, Inches	Actual Inside Diameter, Inches	Actual Outside Diameter, Inches	Circumference, Outside, Inches	Weight per Foot, Pounds
⅛	0.269	0.405	1.27	0.25
¼	0.364	0.540	1.69	0.43
⅜	0.493	0.675	2.12	0.57
½	0.622	0.840	2.65	0.86
¾	0.824	1.050	3.29	1.14
1	1.049	1.315	4.13	1.68
1¼	1.380	1.660	5.21	2.28
1½	1.610	1.900	5.96	2.72
2	2.067	2.375	7.46	3.66
2½	2.469	2.875	9.03	5.80
3	3.068	3.500	10.96	7.58
3½	3.548	4.000	12.56	9.11
4	4.026	4.500	14.13	10.80
5	5.047	5.563	17.47	14.70
6	6.065	6.625	20.81	19.00
8	7.981	8.625	27.09	28.60
10	10.020	10.750	33.77	40.50
12	11.938	12.750	40.05	53.60
14	13.126	14.000	47.12	63.30
16	15.000	16.000	53.41	82.80
18	16.876	18.000	56.55	105.00
20	18.814	20.000	62.83	123.00
24	22.626	24.000	75.40	171.00

Dimensions of Schedule 80 (Extra Strong) Steel and Wrought Iron Pipe

Nominal Diameter, Inches	Actual Inside Diameter, Inches	Actual Outside Diameter, Inches	Circumference, Outside, Inches	Weight per Foot, Pounds
⅛	0.215	0.405	1.27	0.314
¼	0.302	0.540	1.69	0.535
⅜	0.423	0.675	2.12	0.738
½	0.546	0.840	2.65	1.087
¾	0.742	1.050	3.29	1.473
1	0.957	1.315	4.13	2.171
1¼	1.278	1.660	5.21	2.996
1½	1.500	1.900	5.96	3.631
2	1.939	2.375	7.46	5.022
2½	2.323	2.875	9.03	7.661
3	2.900	3.500	10.96	10.252
3½	3.364	4.000	12.56	12.505
4	3.826	4.500	14.13	14.983
5	4.813	5.563	17.47	20.778
6	5.761	6.625	20.81	28.573
8	7.625	8.625	27.09	43.388
10	9.564	10.750	33.77	64.400
12	11.376	12.750	40.05	88.600
14	12.500	14.000	47.12	107.00
16	14.314	16.000	53.41	137.00
18	16.126	18.000	56.55	171.00
20	17.938	20.000	62.83	209.00
24	21.564	24.000	75.40	297.00

Dimensions of Schedule 120
Steel and Wrought Iron Pipe

Nominal Diameter, Inches	Actual Inside Diameter, Inches	Actual Outside Diameter, Inches	Area, Inside, Square Inches	Weight per Foot, Pounds
4	3.626	4.500	10.33	19.0
5	4.563	5.563	16.35	27.1
6	5.501	6.625	23.77	36.4
8	7.189	8.625	40.59	60.7
10	9.064	10.75	64.53	89.2
12	10.750	12.75	90.76	126.0
14	11.876	14.0	110.77	147.0
16	13.564	16.0	144.50	193.0
18	15.314	18.0	184.19	239.0
20	17.000	20.0	226.98	297.0
24	20.500	24.0	330.06	416.0

Notes:

Dimensions of Standard Weight steel pipe correspond to Schedule 40 sizes only from ⅛ through 10 inches. From 12 through 24 inches, wall thickness is constant at 0.375 inches.

Dimensions of Extra Strong steel pipe correspond to Schedule 80 sizes only from ⅛ through 8 inches. From 10 through 24 inches, wall thickness is constant at 0.500 inches.

Double Extra Strong pipe has no corresponding schedule number.

Standard Weight, Extra Strong, and Double Extra Strong wrought iron pipe have thicknesses slightly greater than those of corresponding sizes for steel pipe.

Weights given are for pipe with plain ends.

Dimensions of Stainless Steel Pipe

Nominal Diameter, Inches	Outside Diameter, Inches	Wall Thickness, Inches	
		Schedule 40S	Schedule 80S
⅛	0.405	0.068	0.095
¼	0.540	0.088	0.119
⅜	0.675	0.091	0.126
½	0.840	0.109	0.147
¾	1.050	0.113	0.154
1	1.315	0.133	0.179
1¼	1.660	0.140	0.191
1½	1.900	0.145	0.200
2	2.375	0.154	0.218
2½	2.875	0.203	0.276
3	3.500	0.216	0.300
3½	4.000	0.226	0.318
4	4.500	0.237	0.337
5	5.563	0.258	0.375
6	6.625	0.280	0.432
8	8.625	0.322	0.500
10	10.750	0.365	0.500
12	12.750	0.375	0.500

Dimensions of Type K Copper Tube

Nominal Diameter, Inches	Actual Inside Diameter, Inches	Outside Diameter, Inches	Circumference, Outside, Inches	Weight per Foot, Pounds
⅛	0.186	0.250	0.785	0.085
¼	0.311	0.375	1.178	0.134
⅜	0.402	0.500	1.570	0.269
½	0.527	0.625	1.963	0.344
⅝	0.652	0.750	2.355	0.418
¾	0.745	0.875	2.748	0.641
1	0.995	1.125	3.533	0.839
1¼	1.245	1.375	4.318	1.04
1½	1.481	1.625	5.103	1.36
2	1.959	2.125	6.673	2.06
2½	2.435	2.625	8.243	2.93
3	2.907	3.125	9.813	4.00
4	3.857	4.125	12.953	6.51
5	4.805	5.125	16.093	9.67
6	5.741	6.125	19.233	13.9
8	7.583	8.125	25.513	25.9
10	9.449	10.125	31.793	40.3
12	11.315	12.125	38.073	57.8

Types K and L copper tube is made in both hard and soft temper, Type M in hard temper only. Hard temper tube is furnished in straight lengths and is recommended for straight runs of piping, especially for exposed piping where appearance is a factor.

Selection among the different types of copper tube is made on the basis of the following information furnished by Revere Copper and Brass Inc.

Dimensions of Type L Copper Tube

Nominal Diameter, Inches	Actual Inside Diameter, Inches	Outside Diameter, Inches	Circum-ference, Outside, Inches	Weight per Foot, Pounds
⅛	0.200	0.250	0.785	0.068
¼	0.315	0.375	1.178	0.126
⅜	0.430	0.500	1.570	0.198
½	0.545	0.625	1.963	0.285
⅝	0.666	0.750	2.355	0.362
¾	0.785	0.875	2.748	0.455
1	1.025	1.125	3.533	0.655
1¼	1.265	1.375	4.318	0.884
1½	1.505	1.625	5.103	1.14
2	1.985	2.125	6.673	1.75
2½	1.465	2.625	8.243	2.48
3	2.945	3.125	9.813	3.33
4	3.905	4.125	12.953	5.38
5	4.875	5.125	16.093	7.61
6	5.845	6.125	19.233	10.2
8	7.725	8.125	25.513	19.3
10	9.625	10.125	31.793	30.1
12	11.565	12.125	38.073	40.4

Type K is a heavy wall, Type L a medium and Type M a light wall tube. Types K and L can be easily bent cold, and large radius bends in these types in soft temper can be made by hand. Bending is not recommended for Type L hard temper or for Type M.

Soft temper Type K is recommended for underground lines carrying water, oil or gas. Its thick walls are

Dimensions of Type M Copper Tube

Nominal Diameter, Inches	Actual Inside Diameter, Inches	Outside Diameter, Inches	Circumference, Outside, Inches	Weight per Foot, Pounds
⅛	0.200	0.250	0.785	0.068
¼	0.325	0.375	1.178	0.107
⅜	0.450	0.500	1.570	0.145
½	0.569	0.625	1.963	0.204
⅝	0.690	0.750	2.355	0.263
¾	0.811	0.875	2.748	0.328
1	1.055	1.125	3.533	0.465
1¼	1.291	1.375	4.318	0.682
1½	1.527	1.625	5.103	0.940
2	2.009	2.125	6.673	1.460

adequate to withstand possible damage from backfilling. Typical applications include lawn sprinkling systems, underground farm piping, underground water service lines, and fuel oil lines.

Type L tube is used for general plumbing and heating uses. Type M, the lightest wall, is intended for use with soldered fittings only, and is used to good advantage for vent pipe systems, waste lines, interior drainage, engineered piping systems and other non-pressure applications.

Advantages of copper water tube include the fact that both Types K and L can be bent cold and large radius bends in Types K and L soft temper tube can be made by hand. Convenient, portable benders are available for making necessary short radius bends in tube sizes up to 1 inch, inclusive. Such bends are without wrinkles or flattened portions. They produce streamlining where changes of plane or direction occur.

Dimensions of Type DWV Copper Tube

Nominal Diameter, Inches	Actual Inside Diameter, Inches	Outside Diameter, Inches	Cross-Sectional Area of Opening, Sq. In.	Weight per Foot, Pounds
1¼	1.295	1.375	1.317	0.650
1½	1.541	1.625	1.865	0.809
2	2.041	2.125	3.272	1.07
3	3.035	3.125	7.234	1.69
4	4.009	4.125	12.623	2.87
5	4.981	5.125	19.486	4.43
6	5.959	6.125	27.890	6.10

Copper drainage tube has been developed specifically for sanitary drainage service. It can be installed by a simple soldering operation in which the solder is drawn into the joint by capillary action. A complete selection of drainage fittings, featuring long radius elbows, are available for soldered connection to copper tube. These fittings also include fittings with one or more outlets threaded for use with brass or copper pipe or with slip joints. Bends, both adjustable and fixed, with suitable flanges for closet connections, are also available.

In connection with residences, the actual selection of sizes in most instances is controlled by code requirements, but a 3-inch copper tube stack with not over three branch intervals will accommodate 30 fixture units (not over two water closets). Where not fixed by code, a sanitary engineer or other qualified designer would probably specify a 3-inch stack, which will easily fit between 2 x 4 studs. Copper drainage tube may be used equally well in multi-story buildings. Vent headers and waste lines should be long enough to take up expansion in the stacks.

Threading Pipe

Clean, smooth pipe threads are essential to a good joint and depend largely upon the rake or lip angle and lead of the chasers, and the clearance, chip space and number of chasers in the die-head. The lip angle should vary from 15 to 25 degrees, depending upon the style and condition of the chasers and chaser holders. The chip space in front of the chasers should be large enough to allow room for accumulation of chips and at the same time provide means of lubricating the chasers. This is an important point, as insufficient chip space will cause the chips to clog and tear the threads. The lead of the chaser is the angle which is machined or ground on the leading or front side, to enable the die to start readily on the pipe, and also to distribute the work of cutting over a number of threads. To secure a good thread, the lead should cover the first three threads. As the heaviest cutting is done by this beveled part, it should have a slightly greater clearance angle than the rest of the threads on the chaser. When re-grinding chasers which have become dull on the lead, care should be taken to give each chaser the same length of lead, as otherwise the work will be unevenly distributed.

The number of chasers with which a die should be equipped depends upon the size of the die. The number recommended for different sizes is as follows:

Size of Die	Number of Chasers	Size of Die	Number of Chasers
Up to 1¼ inch	4	10 to 12 inches	12
1¼ to 4 inches	6	12 to 14 inches	14
4 to 7 inches	8	14 to 18 inches	16
7 to 10 inches	10	18 to 20 inches	18

Pipe threading dies should be lubricated with a good quality of lard oil or crude cotton-seed oil, the lubricant being used in liberal quantities.

Standard Threads for Steel Pipe

Nominal Diameter, Inches	Threads per Inch	Tap Drill or Bore Size, Inches
$\frac{1}{8}$	27	$\frac{11}{32}$
$\frac{1}{4}$	18	$\frac{7}{16}$
$\frac{3}{8}$	18	$\frac{37}{64}$
$\frac{1}{2}$	14	$\frac{45}{64}$
$\frac{3}{4}$	14	$\frac{29}{32}$
1	$11\frac{1}{2}$	$1\frac{9}{64}$
$1\frac{1}{4}$	$11\frac{1}{2}$	$1\frac{1}{2}$
$1\frac{1}{2}$	$11\frac{1}{2}$	$1\frac{23}{32}$
2	$11\frac{1}{2}$	$2\frac{3}{16}$
$2\frac{1}{2}$	8	$2\frac{11}{16}$
3	8	$3\frac{5}{16}$
$3\frac{1}{2}$	8	$3\frac{13}{16}$
4	8	$4\frac{5}{16}$
5	8	$5\frac{3}{8}$
6	8	$6\frac{7}{16}$
8	8	$8\frac{7}{16}$
10	8	$10\frac{9}{16}$
12	8	$12\frac{9}{16}$
14	8	$13\frac{13}{16}$
16	8	$15\frac{13}{16}$
18	8	$17\frac{13}{16}$
20	8	$19\frac{13}{16}$
24	8	$23\frac{13}{16}$

Length of Engagement, Standard Pipe Thread

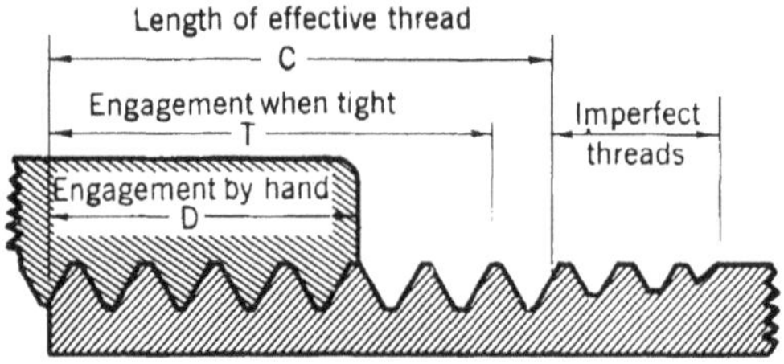

Nominal Diameter, Inches	Length of Thread (C), Inches	Engagement by Hand (D), Inches	Engagement When Tight (T), Inches
1/8	0.26	0.18	1/4
1/4	0.40	0.20	3/8
3/8	0.41	0.24	3/8
1/2	0.53	0.32	1/2
3/4	0.55	0.34	9/16
1	0.68	0.40	11/16
1 1/4	0.71	0.42	11/16
1 1/2	0.72	0.42	11/16
2	0.76	0.44	3/4
2 1/2	1.14	0.68	15/16
3	1.20	0.77	1
3 1/2	1.25	0.82	1 1/16
4	1.30	0.84	1 1/8
5	1.41	0.94	1 1/4
6	1.51	0.96	1 5/16
8	1.71	1.06	1 7/16
10	1.93	1.21	1 5/8
12	2.13	1.36	1 3/4

Overall Dimensions, Cast Iron and Malleable Elbows, Tees and Crosses
(125 lb. Cast Iron and 150 lb. Malleable)

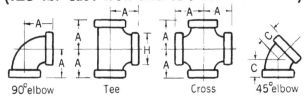

90°elbow   Tee   Cross   45°elbow

Nominal Diameter, Inches	Center-to-End (A) (Except 45° Elbow), Inches	Center-to-End (C) (45° Elbow), Inches	Diameter of Band (H), In.	
			125 lb. Cast Iron	150 lb. Malleable
¼	0.81	0.73	0.93	0.84
⅜	0.95	0.80	1.12	1.02
½	1.12	0.88	1.34	1.20
¾	1.31	0.98	1.63	1.46
1	1.50	1.12	1.95	1.77
1¼	1.75	1.29	2.39	2.15
1½	1.94	1.43	2.68	2.43
2	2.25	1.68	3.28	2.96
2½	2.70	1.95	3.86	3.59
3	3.08	2.17	4.62	4.29
3½	3.42	2.39	5.20	4.84
4	3.79	2.61	5.79	5.40
5	4.50	3.05	7.05	6.58
6	5.13	3.46	8.28	7.77
8	6.56	4.28	10.63	—
10	8.08	5.16	13.12	—
12	9.50	5.97	15.47	—

Overall Dimensions, Cast Iron and Malleable Elbows, Tees and Crosses (250 lb. Iron and 300 lb. Malleable)

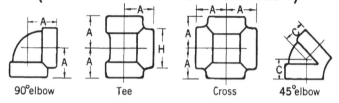

90°elbow   Tee   Cross   45°elbow

Nominal Diameter, Inches	Center-to-End (A) (Except 45° Elbow), Inches	Center-to-End (C) (45° Elbow), Inches	Diameter of Band (H), In.	
			250 lb. Cast Iron	300 lb. Malleable
¼	0.94	0.81	1.17	0.93
⅜	1.06	0.88	1.36	1.12
½	1.25	1.00	1.59	1.34
¾	1.44	1.13	1.88	1.63
1	1.63	1.31	2.24	1.95
1¼	1.94	1.50	2.73	2.39
1½	2.13	1.69	3.07	2.68
2	2.50	2.00	3.74	3.28
2½	2.94	2.25	4.60	3.86
3	3.38	2.50	5.36	4.62
3½	3.75	2.63	5.98	—
4	4.13	2.81	6.61	—
5	4.88	3.19	7.92	—
6	5.63	3.50	9.24	—
8	7.00	4.31	11.73	—
10	8.63	5.19	14.37	—
12	10.00	6.00	16.84	—

Center-to-End Dimensions of Screwed Reducing Elbows

(125 lb. Cast Iron)

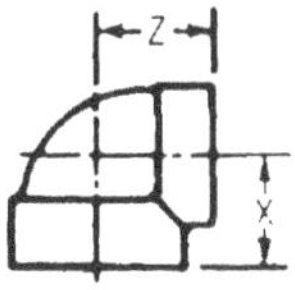

Nominal Diameters, In.	Center-to-End Dimension	
	X, Inches	Z, Inches
½ x ⅜	1 1/16	1
¾ x ½	1 3/16	1¼
1 x ¾	1⅜	1 7/16
1 x ½	1¼	1⅜
1¼ x 1	1 9/16	1 11/16
1¼ x ¾	1 7/16	1⅝
1¼ x ½	1 5/16	1½
1½ x 1¼	1 13/16	1⅞
1½ x 1	1⅝	1 13/16
1½ x ¾	1½	1¾
1½ x ½	1 7/16	1 11/16
2 x 1½	2	2 3/16
2 x 1¼	1⅞	2⅛
2 x 1	1¾	2
2 x ¾	1⅝	2
2 x ½	1½	1⅞
2½ x 2	2⅜	2⅝
2½ x 1½	2 3/16	2½

Center-to-End Dimensions of Screwed Reducing Elbows

(125 lb. Cast Iron, Continued)

Nominal Diameters, In.	Center-to-End Dimension	
	X, Inches	Z, Inches
$2\frac{1}{2}$x$1\frac{1}{4}$	$2\frac{1}{16}$	$2\frac{7}{16}$
$2\frac{1}{2}$x1	$1\frac{7}{8}$	$2\frac{3}{8}$
3 x$2\frac{1}{2}$	$2\frac{13}{16}$	3
3 x2	$2\frac{1}{2}$	$2\frac{7}{8}$
3 x$1\frac{1}{2}$	$2\frac{5}{16}$	$2\frac{13}{16}$
3 x$1\frac{1}{4}$	$2\frac{3}{16}$	$2\frac{3}{4}$
$3\frac{1}{2}$x3	$3\frac{3}{16}$	$3\frac{5}{16}$
4 x$3\frac{1}{2}$	$3\frac{9}{16}$	$3\frac{11}{16}$
4 x3	$3\frac{5}{16}$	$3\frac{5}{8}$
4 x$2\frac{1}{2}$	$3\frac{1}{16}$	$3\frac{1}{2}$
4 x2	$2\frac{3}{4}$	$3\frac{7}{16}$
5 x4	4	$4\frac{7}{16}$
5 x3	$3\frac{1}{2}$	$4\frac{1}{4}$
5 x$2\frac{1}{2}$	$3\frac{1}{4}$	$4\frac{1}{8}$
6 x5	$4\frac{5}{8}$	5
6 x4	$4\frac{1}{8}$	$4\frac{15}{16}$
6 x3	$3\frac{5}{8}$	$4\frac{3}{4}$
8 x6	$5\frac{9}{16}$	$6\frac{3}{8}$

Center-to-End Dimensions of Screwed Reducing Crosses

(125 lb. Cast Iron)

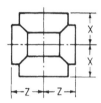

Nominal Diameters, In.	Center-to-End Dimension	
	X, Inches	Z, Inches
$\frac{3}{4}$x $\frac{1}{2}$	$1\frac{3}{16}$	$1\frac{1}{4}$
1 x $\frac{3}{4}$	$1\frac{3}{8}$	$1\frac{7}{16}$
$1\frac{1}{4}$x1	$1\frac{9}{16}$	$1\frac{11}{16}$
$1\frac{1}{4}$x $\frac{3}{4}$	$1\frac{7}{16}$	$1\frac{5}{8}$
$1\frac{1}{2}$x$1\frac{1}{4}$	$1\frac{13}{16}$	$1\frac{7}{8}$
$1\frac{1}{2}$x1	$1\frac{5}{8}$	$1\frac{13}{16}$
$1\frac{1}{2}$x $\frac{3}{4}$	$1\frac{1}{2}$	$1\frac{3}{4}$
2 x$1\frac{1}{2}$	2	$2\frac{3}{16}$
2 x$1\frac{1}{4}$	$1\frac{7}{8}$	$2\frac{1}{8}$
2 x1	$1\frac{3}{4}$	2
2 x $\frac{3}{4}$	$1\frac{5}{8}$	2
$2\frac{1}{2}$x2	$2\frac{3}{8}$	$2\frac{5}{8}$
$2\frac{1}{2}$x$1\frac{1}{2}$	$2\frac{3}{16}$	$2\frac{1}{2}$
$2\frac{1}{2}$x$1\frac{1}{4}$	$2\frac{1}{16}$	$2\frac{7}{16}$
$2\frac{1}{2}$x1	$1\frac{7}{8}$	$2\frac{3}{8}$
3 x2	$2\frac{1}{2}$	$2\frac{7}{8}$
3 x$1\frac{1}{2}$	$2\frac{5}{16}$	$2\frac{13}{16}$

Center-to-End Dimensions of
Screwed Reducing Crosses
(125 lb. Cast Iron, Continued)

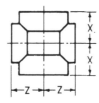

Nominal Diameters, In.	Center-to-End Dimension	
	X, Inches	Z, Inches
3 x1¼	$2\frac{3}{16}$	$2\frac{3}{4}$
3 x1	2	$2\frac{11}{16}$
3½x2½	$2\frac{15}{16}$	$3\frac{1}{4}$
3½x2	$2\frac{5}{8}$	$3\frac{1}{8}$
3½x1½	$2\frac{3}{8}$	$3\frac{1}{16}$
4 x3	$3\frac{5}{16}$	$3\frac{5}{8}$
4 x2½	$3\frac{1}{16}$	$3\frac{1}{2}$
4 x2	$2\frac{3}{4}$	$3\frac{7}{16}$
4 x1½	$2\frac{1}{2}$	$3\frac{5}{16}$
5 x4	4	$4\frac{7}{16}$
5 x3	$3\frac{1}{2}$	$4\frac{1}{4}$
5 x2	$2\frac{15}{16}$	4
6 x4	$4\frac{1}{8}$	$4\frac{15}{16}$
6 x3	$3\frac{5}{8}$	$4\frac{3}{4}$
6 x2½	$3\frac{3}{8}$	$4\frac{11}{16}$
6 x2	$3\frac{1}{16}$	$4\frac{9}{16}$
8 x6	$5\frac{9}{16}$	$6\frac{3}{8}$
8 x4	$4\frac{1}{2}$	$6\frac{3}{16}$

Center-to-End Dimensions of Screwed Reducing Outlet Tees

(125 lb. Cast Iron)

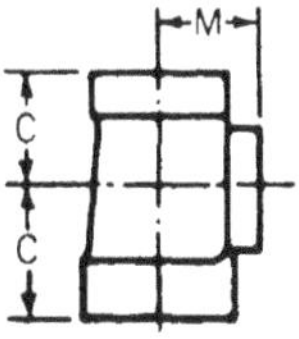

Nominal Diameters, In.	Center-to-End Dimension, Inches	
	Run C	Outlet M
¾x ¾x ⅜	1⅛	1⅛
¾x ¾x ½	1³⁄₁₆	1¼
1 x1 x ⅜	1³⁄₁₆	1¼
1 x1 x ½	1¼	1⅜
1 x1 x ¾	1⅜	1⁷⁄₁₆
1¼x1¼x ½	1⁵⁄₁₆	1½
1¼x1¼x ¾	1⁷⁄₁₆	1⅝
1¼x1¼x1	1⁹⁄₁₆	1¹¹⁄₁₆
1½x1½x ½	1⁷⁄₁₆	1¹¹⁄₁₆
1½x1½x ¾	1½	1¾
1½x1½x1	1⅝	1¹³⁄₁₆
1½x1½x1¼	1¹³⁄₁₆	1⅞
2 x2 x ½	1½	1⅞
2 x2 x ¾	1⅝	2
2 x2 x1	1¾	2
2 x2 x1¼	1⅞	2⅛
2 x2 x1½	2	2³⁄₁₆
2½x2½x ¾	1¾	2⁵⁄₁₆
2½x2½x1	1⅞	2⅜

Center-to-End Dimensions of
Screwed Reducing Outlet Tees
(125 lb. Cast Iron, Continued)

Nominal Diameters, In.	Center-to-End Dimension, Inches	
	Run C	Outlet M
$2\frac{1}{2}$x$2\frac{1}{2}$x$1\frac{1}{4}$	$2\frac{1}{16}$	$2\frac{7}{16}$
$2\frac{1}{2}$x$2\frac{1}{2}$x$1\frac{1}{2}$	$2\frac{3}{16}$	$2\frac{1}{2}$
$2\frac{1}{2}$x$2\frac{1}{2}$x2	$2\frac{3}{8}$	$2\frac{5}{8}$
3 x3 x $\frac{3}{4}$	$1\frac{7}{8}$	$2\frac{5}{8}$
3 x3 x1	2	$2\frac{11}{16}$
3 x3 x$1\frac{1}{4}$	$2\frac{3}{16}$	$2\frac{3}{4}$
3 x3 x$1\frac{1}{2}$	$2\frac{5}{16}$	$2\frac{13}{16}$
3 x3 x2	$2\frac{1}{2}$	$2\frac{7}{8}$
3 x3 x$2\frac{1}{2}$	$2\frac{13}{16}$	3
$3\frac{1}{2}$x$3\frac{1}{2}$x$1\frac{1}{4}$	$2\frac{1}{4}$	3
$3\frac{1}{2}$x$3\frac{1}{2}$x$1\frac{1}{2}$	$2\frac{3}{8}$	$3\frac{1}{16}$
$3\frac{1}{2}$x$3\frac{1}{2}$x2	$2\frac{5}{8}$	$2\frac{1}{8}$
$3\frac{1}{2}$x$3\frac{1}{2}$x$2\frac{1}{2}$	$2\frac{15}{16}$	$3\frac{1}{4}$
$3\frac{1}{2}$x$3\frac{1}{2}$x3	$3\frac{3}{16}$	$3\frac{5}{16}$
4 x4 x1	$2\frac{1}{4}$	$3\frac{3}{16}$
4 x4 x$1\frac{1}{2}$	$2\frac{1}{2}$	$3\frac{3}{16}$
4 x4 x2	$2\frac{3}{4}$	$3\frac{7}{16}$
4 x4 x$2\frac{1}{2}$	$3\frac{1}{16}$	$3\frac{1}{2}$
4 x4 x3	$3\frac{5}{16}$	$3\frac{5}{8}$

Center-to-End Dimensions of Screwed Reducing Outlet Tees

(125 lb. Cast Iron, Continued)

Nominal Diameters, In.	Center-to-End Dimension, Inches	
	Run C	Outlet M
5 x5 x1	$2\frac{7}{16}$	$3\frac{13}{16}$
5 x5 x1¼	$2\frac{5}{8}$	$3\frac{7}{8}$
5 x5 x1½	$2\frac{3}{4}$	$3\frac{15}{16}$
5 x5 x2	$2\frac{15}{16}$	4
5 x5 x2½	$3\frac{1}{4}$	$4\frac{1}{8}$
5 x5 x3	$3\frac{1}{2}$	$4\frac{1}{4}$
5 x5 x4	4	$4\frac{7}{16}$
6 x6 x1½	$2\frac{7}{8}$	$4\frac{1}{2}$
6 x6 x2	$3\frac{1}{16}$	$4\frac{9}{16}$
6 x6 x2½	$3\frac{3}{8}$	$4\frac{11}{16}$
6 x6 x3	$3\frac{5}{8}$	$4\frac{3}{4}$
6 x6 x4	$4\frac{1}{8}$	$4\frac{15}{16}$
6 x6 x5	$4\frac{5}{8}$	5
8 x8 x2	$3\frac{7}{16}$	$5\frac{13}{16}$
8 x8 x2½	$3\frac{11}{16}$	6
8 x8 x3	4	$6\frac{1}{16}$
8 x8 x4	$4\frac{1}{2}$	$6\frac{3}{16}$
8 x8 x5	5	$6\frac{1}{4}$
8 x8 x6	$5\frac{9}{16}$	$6\frac{3}{8}$

Center-to-End Dimensions of Screwed Reducing Elbows

(150 lb. Malleable Iron)

(See drawing next page)

Nominal Diameters, In.	Center-to-End Dimension	
	X, Inches	Z, Inches
$\frac{3}{8}$x $\frac{1}{4}$	$\frac{7}{8}$	$\frac{7}{8}$
$\frac{3}{8}$x $\frac{1}{8}$	$\frac{13}{16}$	$\frac{7}{8}$
$\frac{1}{2}$x $\frac{3}{8}$	$1\frac{1}{16}$	1
$\frac{1}{2}$x $\frac{1}{4}$	1	1
$\frac{3}{4}$x $\frac{1}{2}$	$1\frac{3}{16}$	$1\frac{1}{4}$
$\frac{3}{4}$x $\frac{3}{8}$	$1\frac{1}{8}$	$1\frac{1}{8}$
$\frac{3}{4}$x $\frac{1}{4}$	$1\frac{1}{16}$	$1\frac{1}{16}$
1 x $\frac{3}{4}$	$1\frac{3}{8}$	$1\frac{7}{16}$
1 x $\frac{1}{2}$	$1\frac{1}{4}$	$1\frac{3}{8}$
1 x $\frac{3}{8}$	$1\frac{3}{16}$	$1\frac{1}{4}$
$1\frac{1}{4}$x1	$1\frac{9}{16}$	$1\frac{11}{16}$
$1\frac{1}{4}$x $\frac{3}{4}$	$1\frac{7}{16}$	$1\frac{5}{8}$
$1\frac{1}{4}$x $\frac{1}{2}$	$1\frac{5}{16}$	$1\frac{1}{2}$
$1\frac{1}{2}$x$1\frac{1}{4}$	$1\frac{13}{16}$	$1\frac{7}{8}$
$1\frac{1}{2}$x1	$1\frac{5}{8}$	$1\frac{13}{16}$
$1\frac{1}{2}$x $\frac{3}{4}$	$1\frac{1}{2}$	$1\frac{3}{4}$
2 x$1\frac{1}{2}$	2	$2\frac{3}{16}$
2 x$1\frac{1}{4}$	$1\frac{7}{8}$	$2\frac{1}{8}$
2 x1	$1\frac{3}{4}$	2
2 x $\frac{3}{4}$	$1\frac{5}{8}$	2
$2\frac{1}{2}$x2	$2\frac{3}{8}$	$2\frac{5}{8}$
$2\frac{1}{2}$x$1\frac{1}{2}$	$2\frac{3}{16}$	$2\frac{1}{2}$
3 x$2\frac{1}{2}$	$2\frac{13}{16}$	3
3 x2	$2\frac{1}{2}$	$2\frac{7}{8}$
4 x3	$3\frac{5}{16}$	$3\frac{5}{8}$

Center-to-End Dimensions of Screwed Reducing Crosses
(150 lb. Malleable Iron)

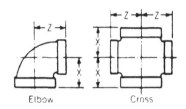

Elbow   Cross

(See table for elbows on previous page)

Nominal Diameters, In.	Center-to-End Dimension	
	X, Inches	Z, Inches
$\frac{3}{4}$x $\frac{1}{2}$	$1\frac{3}{16}$	$1\frac{1}{4}$
1 x $\frac{3}{4}$	$1\frac{3}{8}$	$1\frac{7}{16}$
1 x $\frac{1}{2}$	$1\frac{1}{4}$	$1\frac{3}{8}$
$1\frac{1}{4}$x1	$1\frac{9}{16}$	$1\frac{11}{16}$
$1\frac{1}{4}$x $\frac{3}{4}$	$1\frac{7}{16}$	$1\frac{5}{8}$
$1\frac{1}{2}$x$1\frac{1}{4}$	$1\frac{13}{16}$	$1\frac{7}{8}$
$1\frac{1}{2}$x1	$1\frac{5}{8}$	$1\frac{13}{16}$
$1\frac{1}{2}$x $\frac{3}{4}$	$1\frac{1}{2}$	$1\frac{3}{4}$
2 x$1\frac{1}{2}$	2	$2\frac{3}{16}$
2 x$1\frac{1}{4}$	$1\frac{7}{8}$	$2\frac{1}{8}$
2 x1	$1\frac{3}{4}$	2
2 x $\frac{3}{4}$	$1\frac{5}{8}$	2
$2\frac{1}{2}$x2	$2\frac{3}{8}$	$2\frac{5}{8}$
3 x2	$2\frac{1}{2}$	$2\frac{7}{8}$

Center-to-End Dimensions of
Screwed Reducing Outlet Tees

(150 lb. Malleable Iron)

Nominal Diameters, In.	Center-to-End Dimension, Inches	
	Run C	Outlet M
$\frac{1}{2}$ x $\frac{3}{8}$	$1\frac{1}{16}$	1
$\frac{3}{4}$ x $\frac{1}{2}$	$1\frac{3}{16}$	$1\frac{1}{4}$
$\frac{3}{4}$ x $\frac{3}{8}$	$1\frac{1}{8}$	$1\frac{1}{8}$
$\frac{3}{4}$ x $\frac{1}{4}$	$1\frac{1}{16}$	$1\frac{1}{16}$
1 x $\frac{3}{4}$	$1\frac{3}{8}$	$1\frac{7}{16}$
1 x $\frac{1}{2}$	$1\frac{1}{4}$	$1\frac{3}{8}$
1 x $\frac{3}{8}$	$1\frac{3}{16}$	$1\frac{1}{4}$
1 x $\frac{1}{4}$	$1\frac{1}{8}$	$1\frac{1}{4}$
$1\frac{1}{4}$ x 1	$1\frac{9}{16}$	$1\frac{11}{16}$
$1\frac{1}{4}$ x $\frac{3}{4}$	$1\frac{7}{16}$	$1\frac{5}{8}$
$1\frac{1}{4}$ x $\frac{1}{2}$	$1\frac{5}{16}$	$1\frac{1}{2}$
$1\frac{1}{4}$ x $\frac{3}{8}$	$1\frac{1}{4}$	$1\frac{7}{16}$
$1\frac{1}{2}$ x $1\frac{1}{4}$	$1\frac{13}{16}$	$1\frac{7}{8}$
$1\frac{1}{2}$ x 1	$1\frac{5}{8}$	$1\frac{13}{16}$
$1\frac{1}{2}$ x $\frac{3}{4}$	$1\frac{1}{2}$	$1\frac{3}{4}$
$1\frac{1}{2}$ x $\frac{1}{2}$	$1\frac{7}{16}$	$1\frac{11}{16}$
2 x $1\frac{1}{2}$	2	$2\frac{3}{16}$
2 x $1\frac{1}{4}$	$1\frac{7}{8}$	$2\frac{1}{8}$
2 x 1	$1\frac{3}{4}$	2

Center-to-End Dimensions of Screwed Reducing Outlet Tees

(150 lb. Malleable Iron, Continued)

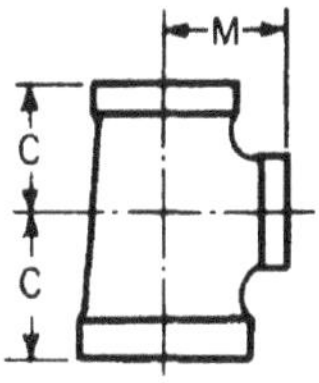

Nominal Diameters, In.	Center-to-End Dimension, Inches	
	Run C	Outlet M
2 x ¾	1⅝	2
2½x2	2⅜	2⅝
2½x1½	2³⁄₁₆	2½
2½x1¼	2¹⁄₁₆	2⁷⁄₁₆
2½x1	1⅞	2⅜
3 x2½	2¹³⁄₁₆	3
3 x2	2½	2⅞
3 x1½	2⁵⁄₁₆	2¹³⁄₁₆
3 x1¼	2³⁄₁₆	2¾
3 x1	2	2¹¹⁄₁₆
4 x3	3⁵⁄₁₆	3⅝
4 x2½	3¹⁄₁₆	3½
4 x2	2¾	3⁷⁄₁₆
4 x1½	2½	3⁵⁄₁₆
5 x3	3½	4¼
6 x4	4⅛	4¹⁵⁄₁₆
6 x3	3⅝	4¾
6 x2½	3⅜	4¹¹⁄₁₆
6 x2	3¹⁄₁₆	4⁹⁄₁₆

Center-to-End Dimensions of Screwed Reducing Elbows

(300 lb. Malleable Iron)

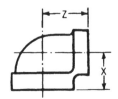

Nominal Diameters, In.	Center-to-End Dimension	
	X, Inches	Z, Inches
½x ⅜	1³⁄₁₆	1³⁄₁₆
¾x ½	1⁵⁄₁₆	1⅜
1 x ¾	1½	1⁹⁄₁₆
1¼x1	1¾	1¹³⁄₁₆
1½x1¼	2	2¹⁄₁₆
2 x1½	2¼	2⅜
2½x2	2¹¹⁄₁₆	2¾
3 x2½	3¹⁄₁₆	3⁵⁄₁₆

End-to-End Dimensions of

Screwed Reducing Couplings

(300 lb. Malleable Iron)

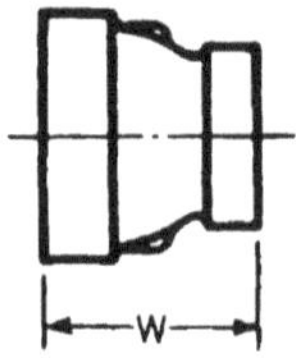

Table is on following page.

End-to-End Dimensions of Screwed Reducing Couplings

(300 lb. Malleable Iron, Continued)

Nominal Diameters, In.	Dimension W, Inches
$\frac{3}{8}$ x $\frac{1}{4}$	$1\frac{7}{16}$
$\frac{1}{2}$ x $\frac{3}{8}$	$1\frac{11}{16}$
$\frac{1}{2}$ x $\frac{1}{4}$	$1\frac{11}{16}$
$\frac{3}{4}$ x $\frac{1}{2}$	$1\frac{3}{4}$
$\frac{3}{4}$ x $\frac{3}{8}$	$1\frac{3}{4}$
$\frac{3}{4}$ x $\frac{1}{4}$	$1\frac{3}{4}$
1 x $\frac{3}{4}$	2
1 x $\frac{1}{2}$	2
1 x $\frac{3}{8}$	2
1 x $\frac{1}{4}$	2
$1\frac{1}{4}$ x 1	$2\frac{3}{8}$
$1\frac{1}{4}$ x $\frac{3}{4}$	$2\frac{3}{8}$
$1\frac{1}{4}$ x $\frac{1}{2}$	$2\frac{3}{8}$
$1\frac{1}{2}$ x $1\frac{1}{4}$	$2\frac{11}{16}$
$1\frac{1}{2}$ x 1	$2\frac{11}{16}$
$1\frac{1}{2}$ x $\frac{3}{4}$	$2\frac{11}{16}$
$1\frac{1}{2}$ x $\frac{1}{2}$	$2\frac{11}{16}$
2 x $1\frac{1}{2}$	$3\frac{3}{16}$
2 x $1\frac{1}{4}$	$3\frac{3}{16}$
2 x 1	$3\frac{3}{16}$
2 x $\frac{3}{4}$	$3\frac{3}{16}$
2 x $\frac{1}{2}$	$3\frac{3}{16}$
$2\frac{1}{2}$ x 2	$3\frac{11}{16}$
$2\frac{1}{2}$ x $1\frac{1}{2}$	$3\frac{11}{16}$
3 x $2\frac{1}{2}$	$4\frac{1}{16}$
3 x 2	$4\frac{1}{16}$
3 x $1\frac{1}{2}$	$4\frac{1}{16}$

Center-to-End Dimensions of Screwed Reducing Outlet Tees

(300 lb. Malleable Iron)

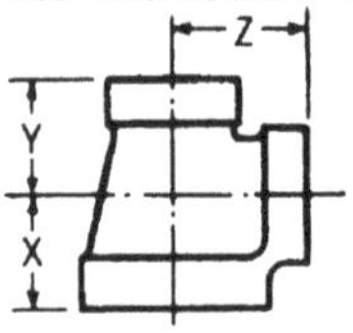

Nominal Diameters, In.	Center-to-End Dimension, Inches	
	Run C	Outlet M
1 x ¾	1½	1⁹⁄₁₆
1 x ½	1⁷⁄₁₆	1½
1 x ⅜	1⁵⁄₁₆	1⁷⁄₁₆
1¼x1	1¾	1¹³⁄₁₆
1¼x ¾	1⅝	1¾
1¼x ½	1½	1¹¹⁄₁₆
1½x1¼	2	2¹⁄₁₆
1½x1	1¹³⁄₁₆	2
1½x ¾	1¹¹⁄₁₆	1⅞
1½x ½	1⅝	1¹³⁄₁₆
2 x1½	2¼	2⅜
2 x1¼	2⅛	2⁵⁄₁₆
2 x1	2	2¼
2 x ¾	1¹³⁄₁₆	2⅛
2 x ½	1¾	2¹⁄₁₆
2½x2	2¹¹⁄₁₆	2¾
2½x1½	2⁷⁄₁₆	2⅝
3 x2½	3¹⁄₁₆	3⁵⁄₁₆
3 x2	2¹³⁄₁₆	3⅛

Length of Pipe Nipples
Long, Short and Close

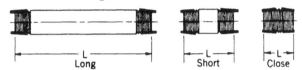

Diameter of Nipples, Inches	Long Nipples, Ranges of Length L*, Inches	Short Nipples, Length L, Inches	Close Nipples, Length L, Inches
$\frac{1}{8}$	2-12	$1\frac{1}{2}$	$\frac{3}{4}$
$\frac{1}{4}$	2-12	$1\frac{1}{2}$	$\frac{7}{8}$
$\frac{3}{8}$	2-12	$1\frac{1}{2}$	1
$\frac{1}{2}$	2-12	$1\frac{1}{2}$	$1\frac{1}{8}$
$\frac{3}{4}$	$2\frac{1}{2}$-12	2	$1\frac{3}{8}$
1	$2\frac{1}{2}$-12	2	$1\frac{1}{2}$
$1\frac{1}{4}$	3-12	$2\frac{1}{2}$	$1\frac{5}{8}$
$1\frac{1}{2}$	3-12	$2\frac{1}{2}$	$1\frac{3}{4}$
2	3-12	$2\frac{1}{2}$	2
$2\frac{1}{2}$	$3\frac{1}{2}$-12	3	$2\frac{1}{2}$
3	$3\frac{1}{2}$-12	3	$2\frac{5}{8}$
$3\frac{1}{2}$	$4\frac{1}{2}$-12	4	$2\frac{3}{4}$
4	$4\frac{1}{2}$-12	4	$2\frac{7}{8}$
5	5-12	$4\frac{1}{2}$	3
6	5-12	$4\frac{1}{2}$	$3\frac{1}{8}$
8	$5\frac{1}{2}$-12	5	$3\frac{1}{2}$
10	8-12	5	$3\frac{7}{8}$
12	8-12	6	$4\frac{1}{2}$

*In $\frac{1}{2}$-inch increments from 2 to 6 inches; in 1-inch increments from 6 to 12 inches.

Clearance Between Pipe and Wall to Provide for Turning Screwed Fittings

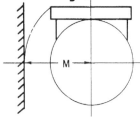

The table below gives the minimum distance from the centerline of a pipe to a wall or other obstruction which will allow standard (125 lb.) cast iron fittings to turn.

Nominal Diameter, Inches	Clearance M, Inches
¼	1
⅜	1⅛
½	1⅜
¾	1⅝
1	1⅞
1¼	2³⁄₁₆
1½	2⁷⁄₁₆
2	2⅞
2½	3⅜
3	3¹⁵⁄₁₆
3½	4⅜
4	4¹³⁄₁₆
5	5¾
6	6⅝
8	8½
10	10⅜
12	12⁵⁄₁₆

Center-to-Center Dimensions of Pipe Lines Provide for Turning of Screwed Fittings

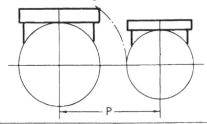

Diameter of One Pipe, Inches	Diameter of Other Pipe, Inches			
	5	6	8	10
	Center-to-Center, P, Inches			
5	$9\frac{5}{16}$	$10\frac{3}{16}$	$12\frac{1}{16}$	$13\frac{7}{8}$
6	$10\frac{3}{16}$	$10\frac{3}{4}$	$12\frac{5}{8}$	$14\frac{1}{2}$
8	$12\frac{1}{16}$	$12\frac{5}{8}$	$13\frac{13}{16}$	$15\frac{11}{16}$
10	$13\frac{7}{8}$	$14\frac{1}{2}$	$15\frac{11}{16}$	$16\frac{15}{16}$
12*	$15\frac{7}{8}$	$16\frac{7}{16}$	$17\frac{5}{8}$	$18\frac{7}{8}$

* Dimension P between two 12 in. lines is 20 inches.

The tables on this page and the three following pages are for determining the minimum distance P between the center lines of two parallel pipe lines to provide for the turning of screwed fittings, assuming that fittings will lie opposite each other. The figures apply to 125 lb. cast iron screwed fittings; dimensions of 150 lb. malleable iron fittings are somewhat less so that the figures given are safe for the latter. The tables provide for the turning of the larger fitting.

Referring to the illustration, it is seen that radius indicated by the dotted line is equal to the diagonal of a triangle of which half the width of the band H is one leg and the dimension A (centerline to edge of band) is the other leg (dimensions A and H are as given in the table of Overall Dimensions, Cast Iron and Malleable

Clearance Between Pipe Lines to Provide for Turning of Screwed Fittings

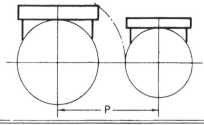

Diameter of One Pipe, Inches	Diameter of Other Pipe, Inches			
	2½	3	3½	4
	Center-to-Center, P, Inches			
2½	5⁵⁄₁₆	5¹³⁄₁₆	6⁵⁄₁₆	6¾
3	5¹³⁄₁₆	6¼	6¹¹⁄₁₆	7⅛
3½	6⁵⁄₁₆	6¹¹⁄₁₆	6¹⁵⁄₁₆	7⁷⁄₁₆
4	6¾	7⅛	7⁷⁄₁₆	7¹¹⁄₁₆
5	7¹¹⁄₁₆	8⅛	8⅜	8¹¹⁄₁₆
6	8⅝	8¹⁵⁄₁₆	9¼	9⁹⁄₁₆
8	10⁷⁄₁₆	10¹³⁄₁₆	11⅛	11⅜
10	12⁵⁄₁₆	12¹¹⁄₁₆	13	13¼
12	14¼	14⅝	14⅞	15³⁄₁₆

Elbows, Tees and Crosses - 125 lb. Cast Iron and 150 lb. Malleable). Consequently, the diagonal or radius is $\sqrt{A^2 + (H/_2)^2}$.

The dimension P is made up of this diagonal plus half the width of the band H of the smaller fitting.

For example, take the case of an 8 inch fitting adjacent to a 5 inch fitting in parallel lines so that clearance must be provided to allow for turning the 8 inch fitting. From the table of overall dimensions, H and A for the 8 inch fitting are 10.63 and 6.56 inches, respectively. H for the 5 inch fitting is 7.05 inches.

Clearance Between Pipe Lines to Provide for Turning of Screwed Fittings

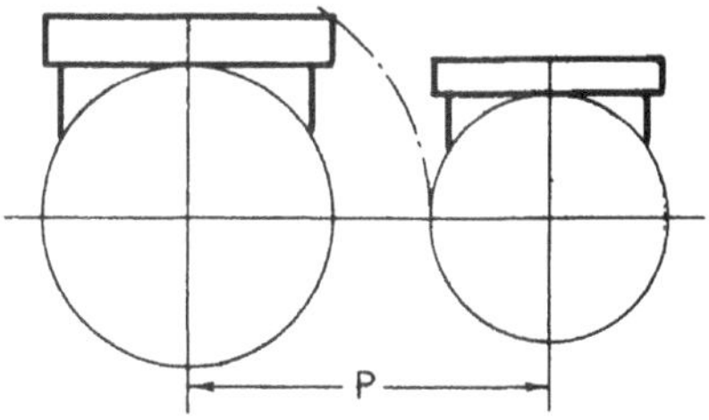

Diameter of One Pipe, Inches	Diameter of Other Pipe, Inches			
	1	1 1/4	1 1/2	2
	Center-to-Center, P, Inches			
1	2 13/16	3 3/16	3 3/8	3 13/16
1 1/4	3 3/16	3 3/8	3 5/8	4 1/16
1 1/2	3 3/8	3 5/8	3 3/4	4 3/16
2	3 13/16	4 1/16	4 3/16	4 1/2
2 1/2	4 3/8	4 9/16	4 11/16	5
3	4 7/8	5 1/8	5 1/4	5 9/16
3 1/2	5 5/16	5 9/16	5 11/16	6
4	5 13/16	6 1/16	6 3/16	6 1/2
5	6 3/4	7	7 1/8	7 7/16
6	7 5/8	7 13/16	8	8 5/16
8	9 7/16	9 11/16	9 13/16	10 1/8
10	11 3/8	11 9/16	11 11/16	12
12	13 5/16	13 1/2	13 5/8	13 15/16

The turning radius for the 8 inch fitting is thus $\sqrt{(6.56)^2 + (5.32)^2}$ or 8.45 inches. Half the width of H for the 5 inch fitting is 3.52. The sum of the two is $8.45 + 3.52 = 11.97$ inches, or 12 inches. The table gives 12 1/16 inches, the 1/16 inch being a margin of safety.

Clearance Between Pipe Lines to Provide for Turning of Screwed Fittings

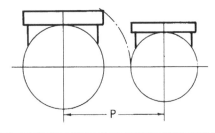

Diameter of One Pipe, Inches	Diameter of Other Pipe, Inches			
	¼	⅜	½	¾
	Center-to-Center, P, Inches			
¼	1½	1¹¹⁄₁₆	1¹³⁄₁₆	2⅛
⅜	1¹¹⁄₁₆	1¾	1¹⁵⁄₁₆	2³⁄₁₆
½	1¹³⁄₁₆	1¹⁵⁄₁₆	2	2¼
¾	2⅛	2³⁄₁₆	2¼	2⁷⁄₁₆
1	2⁵⁄₁₆	2⁷⁄₁₆	2½	2¹¹⁄₁₆
1¼	2⅝	2¹¹⁄₁₆	2⅞	3
1½	2⅞	3	3¹⁄₁₆	3¼
2	3⁵⁄₁₆	3⅜	3½	3¹¹⁄₁₆
2½	3⅞	3¹⁵⁄₁₆	4¹⁄₁₆	4³⁄₁₆
3	4⅜	4½	4⁹⁄₁₆	4¾
3½	4¹³⁄₁₆	4¹⁵⁄₁₆	5	5³⁄₁₆
4	5⁵⁄₁₆	5⅜	5½	5¹¹⁄₁₆
5	6¼	6⅜	6⁷⁄₁₆	6⅝
6	7⅛	7³⁄₁₆	7⁵⁄₁₆	7½
8	9	9¹⁄₁₆	9³⁄₁₆	9⁵⁄₁₆
10	10⅞	10¹⁵⁄₁₆	11¹⁄₁₆	11³⁄₁₆
12	12¾	12⅞	13	13⅛

Laying Lengths, Cast Iron and Malleable 90° Elbows, Tees and Crosses

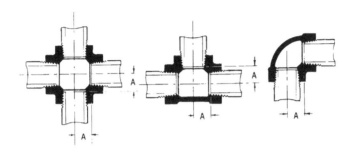

Nominal Diameter, Inches	Dimension A, Inches
¼	⁷⁄₁₆
⅜	⁹⁄₁₆
½	⅝
¾	¾
1	¹³⁄₁₆
1¼	1¹⁄₁₆
1½	1¼
2	1½
2½	1¾
3	2⅛
3½	2⅜
4	2⅝
5	3¼
6	3¹³⁄₁₆
8	5⅛

Laying Lengths of
Ys, Cast Iron and Malleable

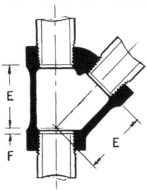

Nominal Diameter, Inches	Dimension E		Dimension F	
	Cast Iron	Malleable	Cast Iron	Malleable
¼	—	—	—	—
⅜	—	1¹⁄₁₆	—	⁵⁄₁₆
½	—	1³⁄₁₆	—	¼
¾	1¹¹⁄₁₆	1½	³⁄₁₆	³⁄₁₆
1	2¹⁄₁₆	1¾	¼	¼
1¼	2⁹⁄₁₆	2¼	⁵⁄₁₆	⁷⁄₁₆
1½	3⅛	2⅝	⅜	½
2	3¾	3¼	½	¹¹⁄₁₆
2½	4¼	3¾	⅝	⅝
3	5⅛	4⁹⁄₁₆	¾	¹¹⁄₁₆
3½	—	—	—	—
4	6½	5¹³⁄₁₆	1	¹³⁄₁₆
5	—	—	—	—
6	—	—	—	—
8	—	—	—	—

Laying Lengths of
45° Elbows, Cast Iron and Malleable

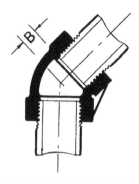

Nominal Diameter, Inches	Dimension B, Inches
$\frac{1}{4}$	$\frac{3}{8}$
$\frac{3}{8}$	$\frac{7}{16}$
$\frac{1}{2}$	$\frac{3}{8}$
$\frac{3}{4}$	$\frac{7}{16}$
1	$\frac{7}{16}$
$1\frac{1}{4}$	$\frac{5}{8}$
$1\frac{1}{2}$	$\frac{3}{4}$
2	$\frac{15}{16}$
$2\frac{1}{2}$	1
3	$1\frac{3}{16}$
$3\frac{1}{2}$	$1\frac{5}{16}$
4	$1\frac{1}{2}$
5	$1\frac{13}{16}$
6	$2\frac{1}{8}$
8	$2\frac{13}{16}$

Laying Lengths of
Reducer Couplings, Cast Iron

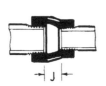 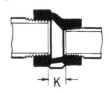

Nominal Diameters, Inches	Dimension J, Inches	Dimension K, Inches
¾ x ½	⁷⁄₁₆	—
1 x ½	½	—
1 x ¾	⅝	—
1¼ x ¾	—	⅞
1¼ x 1	¾	¾
1½ x ¾	—	1
1½ x 1	—	⅞
1½ x 1¼	—	⅞
2 x ¾	—	1⅛
2 x 1	1	1
2 x 1¼	1	1
2 x 1½	1	1
2½ x 1	—	1¹⁄₁₆
2½ x 1¼	—	1¹⁄₁₆
2½ x 1½	—	1¹⁄₁₆
2½ x 2	¹⁵⁄₁₆	1
3 x 1	—	1¼
3 x 1¼	—	1¼
3 x 1½	—	1¼

Laying Lengths of
Reducer Couplings, Cast Iron

Nominal Diameters, Inches	Dimension J, Inches	Dimension K, Inches
3 x2	$1\frac{1}{8}$	$1\frac{3}{16}$
3 x2½	$\frac{15}{16}$	1
3½x1	—	$1\frac{3}{8}$
3½x1¼	—	$1\frac{3}{8}$
3½x1½	—	$1\frac{3}{8}$
3½x2	—	$1\frac{5}{16}$
3½x2½	—	$1\frac{1}{8}$
3½x3	—	$1\frac{1}{16}$
4 x1	—	$1\frac{9}{16}$
4 x1¼	—	$1\frac{9}{16}$
4 x1½	—	$1\frac{9}{16}$
4 x2	$1\frac{1}{2}$	$1\frac{1}{2}$
4 x2½	$1\frac{5}{16}$	$1\frac{5}{16}$
4 x3	$1\frac{1}{4}$	$1\frac{1}{4}$
5 x2	—	$1\frac{7}{8}$
5 x3	—	$1\frac{5}{8}$
5 x4	$1\frac{1}{2}$	$1\frac{1}{2}$
6 x2	—	$2\frac{5}{16}$
6 x3	—	$2\frac{1}{16}$
6 x4	$1\frac{15}{16}$	$1\frac{15}{16}$
6 x5	$1\frac{13}{16}$	$1\frac{13}{16}$
8 x5	—	$2\frac{9}{16}$
8 x6	$2\frac{1}{2}$	$2\frac{1}{2}$

Laying Lengths of
Couplings and Short Nipples, Malleable Iron

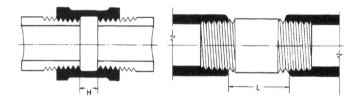

Nominal Diameter, Inches	Dimension H, Coupling, Inches	Dimension L, Short Nipple, Inches
$\frac{1}{8}$	$\frac{7}{16}$	$\frac{1}{4}$
$\frac{1}{4}$	$\frac{5}{16}$	$\frac{1}{8}$
$\frac{3}{8}$	$\frac{7}{16}$	$\frac{1}{4}$
$\frac{1}{2}$	$\frac{5}{16}$	$\frac{1}{8}$
$\frac{3}{4}$	$\frac{3}{8}$	$\frac{1}{4}$
1	$\frac{5}{16}$	$\frac{1}{8}$
$1\frac{1}{4}$	$\frac{9}{16}$	$\frac{1}{4}$
$1\frac{1}{2}$	$\frac{3}{4}$	$\frac{3}{8}$
2	1	$\frac{1}{2}$
$2\frac{1}{2}$	1	$\frac{5}{8}$
3	$1\frac{3}{16}$	$\frac{5}{8}$
$3\frac{1}{2}$	—	$\frac{5}{8}$
4	$1\frac{7}{16}$	$\frac{5}{8}$
5	—	$\frac{1}{2}$
6	—	$\frac{1}{2}$

Laying Lengths of
45° and 90° Malleable Iron Street Elbows

Nominal Diameter, Inches	Dimension A, 90° Elbow, Inches	Dimension C, 45° Elbow, Inches
$\frac{1}{4}$	$\frac{7}{16}$	$\frac{1}{4}$
$\frac{3}{8}$	$\frac{9}{16}$	$\frac{5}{16}$
$\frac{1}{2}$	$\frac{5}{8}$	$\frac{5}{16}$
$\frac{3}{4}$	$\frac{3}{4}$	$\frac{3}{8}$
1	$\frac{13}{16}$	$\frac{3}{8}$
$1\frac{1}{4}$	$1\frac{1}{16}$	$\frac{9}{16}$
$1\frac{1}{2}$	$1\frac{1}{4}$	$\frac{11}{16}$
2	$1\frac{1}{2}$	$\frac{15}{16}$
$2\frac{1}{2}$	$1\frac{3}{4}$	—
3	$2\frac{1}{8}$	—
$3\frac{1}{2}$	$2\frac{3}{8}$	—
4	$2\frac{5}{8}$	—
5	$3\frac{1}{4}$	—
6	$3\frac{13}{16}$	—

Laying Lengths of
Reducer Couplings, Malleable Iron

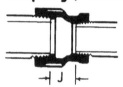

Nominal Diameters, In.	Dimension J, In.	Nominal Diameters, In.	Dimension J, In.
$\frac{3}{8}$x $\frac{1}{4}$	$\frac{3}{8}$	2 x1	$1\frac{3}{8}$
$\frac{1}{2}$x $\frac{1}{8}$	$\frac{1}{2}$	2 x1$\frac{1}{4}$	$1\frac{3}{8}$
$\frac{1}{2}$x $\frac{1}{4}$	$\frac{3}{8}$	2 x1$\frac{1}{2}$	$1\frac{3}{8}$
$\frac{1}{2}$x $\frac{1}{2}$	$\frac{3}{8}$	2$\frac{1}{2}$x1	$1\frac{5}{8}$
$\frac{3}{4}$x $\frac{1}{4}$	$\frac{1}{2}$	2$\frac{1}{2}$x1$\frac{1}{4}$	$1\frac{5}{8}$
$\frac{3}{4}$x $\frac{3}{8}$	$\frac{1}{2}$	2$\frac{1}{2}$x1$\frac{1}{2}$	$1\frac{5}{8}$
$\frac{3}{4}$x $\frac{1}{2}$	$\frac{3}{8}$	2$\frac{1}{2}$x2	$1\frac{9}{16}$
1 x $\frac{1}{4}$	$\frac{5}{8}$	3 x1	2
1 x $\frac{3}{8}$	$\frac{5}{8}$	3 x1$\frac{1}{4}$	2
1 x $\frac{1}{2}$	$\frac{1}{2}$	3 x1$\frac{1}{2}$	2
1 x $\frac{3}{4}$	$\frac{7}{16}$	3 x2	$1\frac{15}{16}$
1$\frac{1}{4}$x $\frac{1}{2}$	$\frac{7}{8}$	3 x2$\frac{1}{2}$	1
1$\frac{1}{4}$x $\frac{3}{4}$	$\frac{13}{16}$	3$\frac{1}{2}$x2	$2\frac{3}{16}$
1$\frac{1}{4}$x1	$\frac{11}{16}$	3$\frac{1}{2}$x3	$1\frac{15}{16}$
1$\frac{1}{2}$x $\frac{1}{2}$	$1\frac{1}{8}$	4 x1$\frac{1}{2}$	$2\frac{9}{16}$
1$\frac{1}{2}$x $\frac{3}{4}$	$1\frac{1}{16}$	4 x2	$2\frac{1}{2}$
1$\frac{1}{2}$x1	$\frac{15}{16}$	4 x2$\frac{1}{2}$	$2\frac{5}{16}$
1$\frac{1}{2}$x1$\frac{1}{4}$	$\frac{15}{16}$	4 x3	$2\frac{1}{4}$
2 x $\frac{1}{2}$	$1\frac{9}{16}$	5 x4	$2\frac{3}{4}$
2 x $\frac{3}{4}$	$1\frac{1}{2}$	6 x4	$1\frac{15}{16}$

Laying Lengths of
Unions and Union Fittings, Malleable Iron

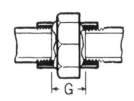

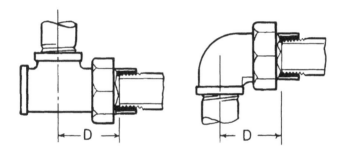

Nominal Diameter, Inches	Dimension G, Union, Inches	Dimension D, Tee or Elbow Union, Inches
$\frac{1}{4}$	1	$1\frac{5}{16}$
$\frac{3}{8}$	$1\frac{1}{8}$	$1\frac{1}{2}$
$\frac{1}{2}$	$1\frac{3}{16}$	$1\frac{5}{8}$
$\frac{3}{4}$	$1\frac{5}{16}$	$1\frac{7}{8}$
1	$1\frac{9}{16}$	$2\frac{3}{16}$
$1\frac{1}{4}$	$1\frac{5}{8}$	$2\frac{1}{2}$
$1\frac{1}{2}$	$1\frac{11}{16}$	$2\frac{11}{16}$
2	$1\frac{13}{16}$	$3\frac{1}{8}$
$2\frac{1}{2}$	$2\frac{5}{16}$	—
3	$2\frac{9}{16}$	—

Laying Lengths of Combined
Tees and Street Elbows, Malleable Iron

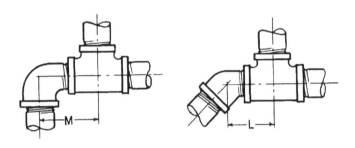

Nominal Diameter, Inches	Dimension M, Inches	Dimension L, Inches
$\frac{1}{8}$	$1\frac{7}{16}$	$1\frac{1}{4}$
$\frac{1}{4}$	$1\frac{5}{8}$	$1\frac{3}{8}$
$\frac{3}{8}$	2	$1\frac{5}{8}$
$\frac{1}{2}$	$2\frac{1}{4}$	$1\frac{13}{16}$
$\frac{3}{4}$	$2\frac{5}{8}$	$2\frac{1}{16}$
1	$2\frac{15}{16}$	$2\frac{5}{16}$
$1\frac{1}{4}$	$3\frac{1}{2}$	$2\frac{3}{4}$
$1\frac{1}{2}$	$3\frac{15}{16}$	$3\frac{1}{8}$
2	$4\frac{3}{4}$	$3\frac{3}{4}$
$2\frac{1}{2}$	$5\frac{9}{16}$	—
3	$6\frac{5}{8}$	—
$3\frac{1}{2}$	—	—
4	$8\frac{5}{16}$	—

Laying Lengths of Standard and Extra Strong Butt Welding Elbows

Long radius

Short radius

45° Long radius

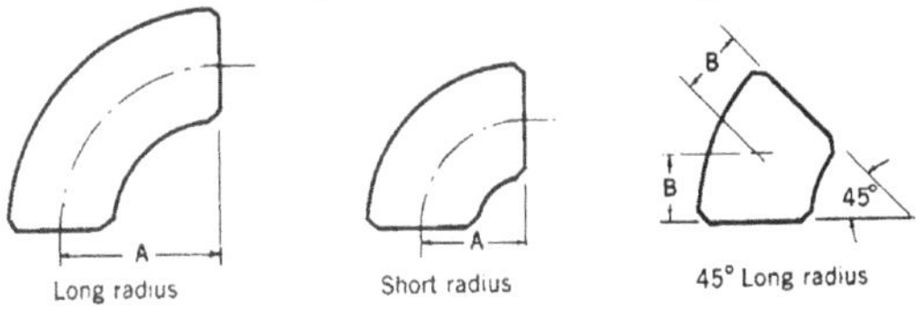

Nominal Diameter, Inches	Dimension A, Inches		45° Elbow Dimension B
	Long Radius 90° Elbow	Short Radius 90° Elbow	
¾	1⅛	—	⁷⁄₁₆
1	1½	1	⅞
1¼	1⅞	1¼	1
1½	2¼	1½	1⅛
2	3	2	1⅜
2½	3¾	2½	1¾
3	4½	3	2
3½	5¼	3½	2¼
4	6	4	2½
5	7½	5	3⅛
6	9	6	3¾
8	12	8	5
10	15	10	6¼
12	18	12	7½
14 O.D.	21	14	8¾
16 O.D.	24	16	10
18 O.D.	27	18	11¼
20 O.D.	30	20	12½
24 O.D.	36	24	15

Laying Lengths of Standard and Extra Strong Butt Welding Straight Tees

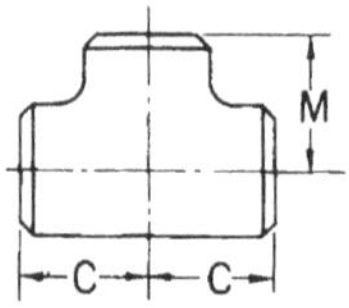

Nominal Diameter, In.	Center-to-End Dimensions, Inches	
	Run C	Outlet M
¾	1⅛	1⅛
1	1½	1½
1¼	1⅞	1⅞
1½	2¼	2¼
2	2½	2½
2½	3	3
3	3⅜	3⅜
3½	3¾	3¾
4	4⅛	4⅛
5	4⅞	4⅞
6	5⅝	5⅝
8	7	7
10	8½	8½
12	10	10
14 O.D.	11	11
16 O.D.	12	12
18 O.D.	13½	13½
20 O.D.	15	15
24 O.D.	17	17

Laying Lengths of
Butt Welding Reducing Outlet Tees

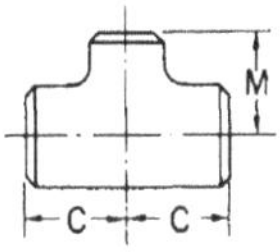

Nominal Diameters, In.	Center-to-End Dimension, Inches	
	Run C	Outlet M
½x ½x ¼	1	1
½x ½x ⅜	1	1
¾x ¾x ⅜	1⅛	1⅛
¾x ¾x ½	1⅛	1⅛
1 x1 x ⅜	1½	1½
1 x1 x ½	1½	1½
1 x1 x ¾	1½	1½
1¼x1¼x ½	1⅞	1⅞
1¼x1¼x ¾	1⅞	1⅞
1¼x1¼x1	1⅞	1⅞
1½x1½x ½	2¼	2¼
1½x1½x ¾	2¼	2¼
1½x1½x1	2¼	2¼
1½x1½x1¼	2¼	2¼
2 x2 x ¾	2½	1¾
2 x2 x1	2½	2
2 x2 x1¼	2½	2¼
2 x2 x1½	2½	2⅜
2½x2½x1	3	2¼
2½x2½x1¼	3	2½

Laying Lengths of
Butt Welding Reducing Outlet Tees

Nominal Diameters, In.	Center-to-End Dimension, Inches	
	Run C	Outlet M
$2\frac{1}{2}$x$2\frac{1}{2}$x$1\frac{1}{2}$	3	$2\frac{5}{8}$
$2\frac{1}{2}$x$2\frac{1}{2}$x2	3	$2\frac{3}{4}$
3 x3 x1	$3\frac{3}{8}$	$2\frac{5}{8}$
3 x3 x$1\frac{1}{4}$	$3\frac{3}{8}$	$2\frac{3}{4}$
3 x3 x$1\frac{1}{2}$	$3\frac{3}{8}$	$2\frac{7}{8}$
3 x3 x2	$3\frac{3}{8}$	3
3 x3 x$2\frac{1}{2}$	$3\frac{3}{8}$	$3\frac{1}{4}$
$3\frac{1}{2}$x$3\frac{1}{2}$x$1\frac{1}{2}$	$3\frac{3}{4}$	$3\frac{1}{8}$
$3\frac{1}{2}$x$3\frac{1}{2}$x2	$3\frac{3}{4}$	$3\frac{1}{4}$
$3\frac{1}{2}$x$3\frac{1}{2}$x$2\frac{1}{2}$	$3\frac{3}{4}$	$3\frac{1}{2}$
$3\frac{1}{2}$x$3\frac{1}{2}$x3	$3\frac{3}{4}$	$3\frac{5}{8}$
4 x4 x$1\frac{1}{2}$	$4\frac{1}{8}$	$3\frac{3}{8}$
4 x4 x2	$4\frac{1}{8}$	$3\frac{1}{2}$
4 x4 x$2\frac{1}{2}$	$4\frac{1}{8}$	$3\frac{3}{4}$
4 x4 x3	$4\frac{1}{8}$	$3\frac{7}{8}$
4 x4 x$3\frac{1}{2}$	$4\frac{1}{8}$	4
5 x5 x2	$4\frac{7}{8}$	$4\frac{1}{8}$
5 x5 x$2\frac{1}{2}$	$4\frac{7}{8}$	$4\frac{1}{4}$
5 x5 x3	$4\frac{7}{8}$	$4\frac{3}{8}$
5 x5 x$3\frac{1}{2}$	$4\frac{7}{8}$	$4\frac{1}{2}$
5 x5 x4	$4\frac{7}{8}$	$4\frac{5}{8}$
6 x6 x$2\frac{1}{2}$	$5\frac{5}{8}$	$4\frac{3}{4}$

Laying Lengths of
Butt Welding Reducing Outlet Tees

| Nominal | Center-to-End Dimension, Inches | |
Diameters, In.	Run C	Outlet M
5x 5x 4	$4\frac{7}{8}$	$4\frac{5}{8}$
6x 6x $2\frac{1}{2}$	$5\frac{5}{8}$	$4\frac{3}{4}$
6x 6x 3	$5\frac{5}{8}$	$4\frac{7}{8}$
6x 6x $3\frac{1}{2}$	$5\frac{5}{8}$	5
6x 6x 4	$5\frac{5}{8}$	$5\frac{1}{8}$
6x 6x 5	$5\frac{5}{8}$	$5\frac{3}{8}$
8x 8x 3	7	6
8x 8x $3\frac{1}{2}$	7	6
8x 8x 4	7	$6\frac{1}{8}$
8x 8x 5	7	$6\frac{3}{8}$
8x 8x 6	7	$6\frac{5}{8}$
10x10x 4	$8\frac{1}{2}$	$7\frac{1}{4}$
10x10x 5	$8\frac{1}{2}$	$7\frac{1}{2}$
10x10x 6	$8\frac{1}{2}$	$7\frac{5}{8}$
10x10x 8	$8\frac{1}{2}$	8
12x12x 5	10	$8\frac{1}{2}$
12x12x 6	10	$8\frac{5}{8}$
12x12x 8	10	9
12x12x10	10	$9\frac{1}{2}$

Laying Lengths of
90° Reducing Elbows

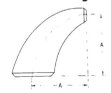

Nominal Diameters, Inches	Dimension A, Inches
2 x1½	3
2 x1	3
2½x2	3¾
2½x1¼	3¾
3 x2½	4½
3 x2	4½
3 x1½	4½
3½x3	5¼
3½x2	5¼
4 x3½	6
4 x3	6
4 x2	6
5 x4	7½
5 x3½	7½
5 x3	7½
5 x2½	7½
6 x5	9
6 x4	9
6 x3½	9
6 x3	9

Laying Lengths of
Butt Welding Reducers

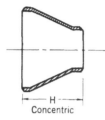
Concentric

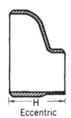
Eccentric

Nominal Diameters, Inches	Dimension H, Inches	Nominal Diameters, Inches	Dimension H, Inches
1 x ⅜	2	3 x1¼	3½
1 x ½	2	3 x1½	3½
1 x ¾	2	3 x2	3½
1¼x ½	2	3 x2½	3½
1¼x ¾	2	3½x1¼	4
1¼x1	2	3½x1½	4
1½x ½	2½	3½x2	4
1½x ¾	2½	3½x3	4
1½x1	2½	4 x1½	4
1½x1¼	2½	4 x2	4
2 x ¾	3	4 x2½	4
2 x1	3	4 x3	4
2 x1¼	3	4 x3½	4
2 x1½	3	5 x2	5
2½x1	3½	5 x2½	5
2½x1¼	3½	5 x3	5
2½x1½	3½	5 x3½	5
2½x2	3½	5 x4	5

Laying Lengths of
Butt Welding Reducers

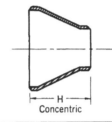
Concentric

Eccentric

Nominal Diameters, Inches	Dimension H, Inches	Nominal Diameters, Inches	Dimension H, Inches
6x 2½	5½	14x 8	13
6x 3	5½	14x10	13
6x 3½	5½	14x12	13
6x 4	5½	16x 8	14
6x 5	5½	16x10	14
8x 3½	6	16x12	14
8x 4	6	16x14	14
8x 5	6	18x10	15
8x 6	6	18x12	15
10x 4	7	18x14	15
10x 5	7	18x16	15
10x 6	7	20x12	20
10x 8	7	20x14	20
12x 5	8	20x16	20
12x 6	8	20x18	20
12x 8	8	24x16	20
12x10	8	24x18	20
14x 6	13	24x20	20

Laying Lengths of
Butt Welding 180° Returns

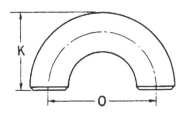

Nominal Diameter, Inches	Long Radius		Short Radius	
	K, Inches	O, Inches	K, Inches	O, Inches
½	1⅞	3	—	—
¾	1¹¹⁄₁₆	2¼	—	—
1	2³⁄₁₆	3	1⅝	2
1¼	2¾	3¾	2¹⁄₁₆	2½
1½	3¼	4½	2⁷⁄₁₆	3
2	4³⁄₁₆	6	3³⁄₁₆	4
2½	5³⁄₁₆	7½	3¹⁵⁄₁₆	5
3	6¼	9	4¾	6
3½	7¼	10½	5½	7
4	8¼	12	6¼	8
5	10⁵⁄₁₆	15	7¾	10
6	12⁵⁄₁₆	18	9⁵⁄₁₆	12
8	16⁵⁄₁₆	24	12⁵⁄₁₆	16
10	20⅜	30	15⅜	20
12	24⅜	36	18⅜	24
14 O.D.	28	42	21	28

Laying Lengths of
Butt Welding Lap Joint Stub Ends

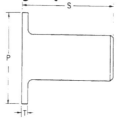

Nominal Diameter, Inches	Diameter P, Inches	Length S, Inches	Wall and Lap Thickness T, Inches
½	1⅜	3	0.109
¾	1¹¹⁄₁₆	3	0.113
1	2	4	0.133
1¼	2½	4	0.140
1½	2⅞	4	0.145
2	3⅝	6	0.154
2½	4⅛	6	0.203
3	5	6	0.216
3½	5½	6	0.226
4	6³⁄₁₆	6	0.237
5	7⁵⁄₁₆	8	0.258
6	8½	8	0.280
8	10⅝	8	0.322
10	12¾	10	0.365
12	15	10	0.375
14 O.D.	16¼	12	0.375
16 O.D.	18½	12	0.375
18 O.D.	21	12	0.375

Laying Lengths of Standard and Extra Strong Butt Welding Caps

Nominal Diameter, Inches	Dimension E, Inches
1	1½
1¼	1½
1½	1½
2	1½
2½	1½
3	2
3 ½	2½
4	2½
5	3
6	3½
8	4
10	5
12	6
14	6½
16	7
18	8
20	9
24	10½

Laying Lengths of Standard and Extra Strong Butt Welding Crosses

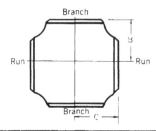

Nominal Diameters, Inches*	Center-to-End of Run, C, Inches	Center-to-End of Branch, B, Inches
1¼x ¾	1⅞	1⅞
1¼x1	1⅞	1⅞
1¼x1¼	1⅞	1⅞
1½x ¾	2¼	2¼
1½x1	2¼	2¼
1½x1¼	2¼	2¼
1½x1½	2¼	2¼
2 x1	2½	2
2 x1¼	2½	2¼
2 x1½	2½	2⅜
2 x2	2½	2½
2½x1	3	2¼
2½x1¼	3	2½
2½x1½	3	2⅝
2½x2	3	2¾
2½x2½	3	3

* First size is run, second is branch diameter.

Laying Lengths of
Standard and Extra Strong
Butt Welding Crosses

Nominal Diameters, Inches	Center-to-End of Run, C, Inches	Center-to-End of Branch, B, inches
3 x1	$3\frac{3}{8}$	$2\frac{5}{8}$
3 x1¼	$3\frac{3}{8}$	$2\frac{3}{4}$
3 x1½	$3\frac{3}{8}$	$2\frac{7}{8}$
3 x2	$3\frac{3}{8}$	3
3 x2½	$3\frac{3}{8}$	$3\frac{1}{4}$
3 x3	$3\frac{3}{8}$	$3\frac{3}{8}$
3½x1½	$3\frac{3}{4}$	$3\frac{1}{8}$
3½x2	$3\frac{3}{4}$	$3\frac{1}{4}$
3½x2½	$3\frac{3}{4}$	$3\frac{1}{2}$
3½x3	$3\frac{3}{4}$	$3\frac{5}{8}$
3½x3½	$3\frac{3}{4}$	$3\frac{3}{4}$
4 x1½	$4\frac{1}{8}$	$3\frac{3}{8}$
4 x2	$4\frac{1}{8}$	$3\frac{1}{2}$
4 x2½	$4\frac{1}{8}$	$3\frac{3}{4}$
4 x3	$4\frac{1}{8}$	$3\frac{7}{8}$
4 x3½	$4\frac{1}{8}$	4
4 x4	$4\frac{1}{8}$	$4\frac{1}{8}$
5 x2	$4\frac{7}{8}$	$4\frac{1}{8}$
5 x2½	$4\frac{7}{8}$	$4\frac{1}{4}$
5 x3	$4\frac{7}{8}$	$4\frac{3}{8}$
5 x3½	$4\frac{7}{8}$	$4\frac{1}{2}$
5 x4	$4\frac{7}{8}$	$4\frac{5}{8}$
5 x5	$4\frac{7}{8}$	$4\frac{7}{8}$

Laying Lengths of Standard and Extra Strong Butt Welding Crosses

Nominal Diameters, Inches	Center-to-End of Run, C, Inches	Center-to-End of Branch, B, Inches
6x 2½	5⅝	4¾
6x 3	5⅝	4⅞
6x 3½	5⅝	5
6x 4	5⅝	5⅛
6x 5	5⅝	5⅜
6x 6	5⅝	5⅝
8x 3	7	6
8x 3½	7	6
8x 4	7	6⅛
8x 5	7	6⅜
8x 6	7	6⅝
8x 8	7	7
10x 4	8½	7¼
10x 5	8½	7½
10x 6	8½	7⅝
10x 8	8½	8
10x10	8½	8½
12x 5	10	8½
12x 6	10	8⅝
12x 8	10	9
12x10	10	9½
12x12	10	10

Laying Lengths of Solid Wedge and Double Disc Steel Gate Valves with Butt Welding Ends

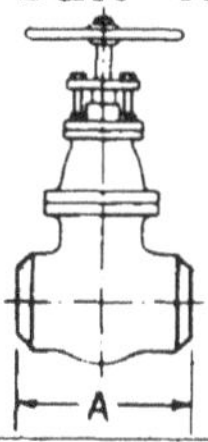

Nominal Size, Inches	Pressure, Lb. per Sq. In.			
	150	300	400	600
	Dimension A, Inches			
1	5	—	8½	8½
1¼	5½	—	9	9
1½	6½	7½	9½	9½
2	8½	8½	11½	11½
2½	9½	9½	13	13
3	11⅛	11⅛	14	14
4	12	12	16	17
5	15	15	18	20
6	15⅞	15⅞	19½	22
8	16½	16½	23½	26
10	18	18	26½	31
12	19¾	19¾	30	33
14	22½	30	32½	35
16	24	33	35½	39
18	26	36	38½	43
20	28	39	41½	47
24	32	45	48½	55

Laying Lengths of
Solid Wedge and Double Disc Steel Gate
Valves with Butt Welding Ends

Nominal Size, Inches	Pressure, Lb. per Sq. In.		
	900	1500	2500
	Dimension A, Inches		
1	10	10	12⅛
1¼	11	11	13¾
1½	12	12	15⅛
2	14½	14½	17¾
2½	16½	16½	20
3	15	18½	22¾
4	18	21½	26½
5	22	26½	31¼
6	24	27¾	36
8	29	32¾	40¼
10	33	39	50
12	38	44½	56
14	40½	49½	—
16	44½	54½	—
18	48	60½	—
20	52	65½	—
24	61	76½	—

Laying Lengths of
Steel Globe and Angle Valves
With Butt Welding Ends

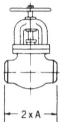
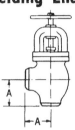

Nominal Size, Inches	Pressure, Lb. per Sq. In.			
	150	300	400	600
	Dimension 2 x A, Inches			
½	4¼	6	6½	6½
¾	4⅝	7	7½	7½
1	5	8	8½	8½
1¼	5½	8½	9	9
1½	6½	9	9½	9½
2	8	10½	11½	11½
2½	8½	11½	13	13
3	9½	12½	14	14
4	11½	14	16	17
5	14	15¾	18	20
6	16	17½	19½	22
8	19½	22	23½	26
10	24½	24½	26½	31
12	27½	28	30	33
14	31	—	—	—
16	36	—	—	—

Laying Lengths of
Steel Globe and Angle Valves
With Butt Welding Ends

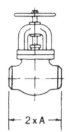
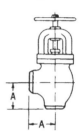

Nominal Size Inches	Pressure, Lb. per Sq. In.		
	900	**1500**	**2500**
	Dimension 2 x A, Inches		
½	—	—	10⅜
¾	9	9	10¾
1	10	10	12⅛
1¼	11	11	13¾
1½	12	12	15⅛
2	14½	14½	17¾
2½	16½	16½	20
3	15	18½	22¾
4	18	21½	26½
5	22	26½	31¼
6	24	27¾	36
8	29	32¾	40¼
10	33	39	50
12	38	44½	56
14	40½	49½	—

Laying Lengths of
Steel Swing Check Valves
With Butt Welding Ends

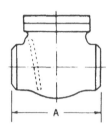

Nominal Size, Inches	Pressure, Lb. per Sq. In.			
	150	300	400	600
	Dimension A, Inches			
½	4¼	—	6½	6½
¾	4⅝	—	7½	7½
1	5	8½	8½	8½
1¼	5½	9	9	9
1½	6½	9½	9½	9½
2	8	10½	11½	11½
2½	8½	11½	13	13
3	9½	12½	14	14
4	11½	14	16	17
5	13	15¾	18	20
6	14	17½	19½	22
8	19½	21	23½	26
10	24½	24½	26½	31
12	27½	28	30	33
14	31	—	—	—

Laying Lengths of
Steel Swing Check Valves
With Butt Welding Ends

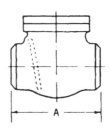

Nominal Size, Inches	Pressure, Lb. per Sq. In.		
	900	1500	2500
	Dimension A, Inches		
$\frac{1}{2}$	—	—	$10\frac{3}{8}$
$\frac{3}{4}$	9	9	$10\frac{3}{4}$
1	10	10	$12\frac{1}{8}$
$1\frac{1}{4}$	11	11	$13\frac{3}{4}$
$1\frac{1}{2}$	12	12	$15\frac{1}{8}$
2	$14\frac{1}{2}$	$14\frac{1}{2}$	$17\frac{3}{4}$
$2\frac{1}{2}$	$16\frac{1}{2}$	$16\frac{1}{2}$	20
3	15	$18\frac{1}{2}$	$22\frac{3}{4}$
4	18	$21\frac{1}{2}$	$26\frac{1}{2}$
5	22	$26\frac{1}{2}$	$31\frac{1}{4}$
6	24	$27\frac{3}{4}$	36
8	29	$32\frac{3}{4}$	$40\frac{1}{4}$
10	33	39	50
12	38	$44\frac{1}{2}$	56
14	$40\frac{1}{2}$	$49\frac{1}{2}$	—

Length of Chords for Spacing
Off Circumference of Circles

The table below gives the length of chord when the diameter is 1. For other diameters, multiply the diameter by the length in the table.

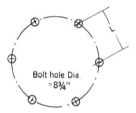

Bolt hole Dia.
= 8¾"

EXAMPLE

A blind flange has a bolt hole diameter of 8¾ inches. Six bolts are to be drilled. What is the spacing L between each hole?

SOLUTION

In the table below, for 6 spaces find the value to be 0.5000. Multiply this by 8¾ inches and obtain 4.375 or 4⅜ inches, the length of the chord, or setting of the dividers in spacing off the holes.

No. of Spaces	Length of Chord	No. of Spaces	Length of Chord
3	0.866025	16	0.195090
4	0.707106	17	0.183749
5	0.587785	18	0.173648
6	0.500000	19	0.164594
7	0.433883	20	0.156434
8	0.382683	21	0.149042
9	0.342020	22	0.142314
10	0.309017	23	0.136166
11	0.281732	24	0.130526
12	0.258819	25	0.125333
13	0.239315	26	0.120536
14	0.222520	27	0.116092
15	0.207911	28	0.111964

Templates for Drilling 25-Lb.
Cast Iron Flanged Fittings

Pipe Diameter, Inches	Flange Diameter, Inches	Flange Thickness, Inches	Bolt Circle Diameter, Inches	Number of Bolts
4	9	¾	7½	8
5	10	¾	8½	8
6	11	¾	9½	8
8	13½	¾	11¾	8
10	16	⅞	14¼	12
12	19	1	17	12
14	21	1⅛	18¾	12
16	23½	1⅛	21¼	16
18	25	1¼	22¾	16
20	27½	1¼	25	20
24	32	1⅜	29½	20
30	38¾	1½	36	28
36	46	1⅝	42¾	32
42	53	1¾	49½	36
48	59½	2	56	44
54	66¼	2¼	62¾	44
60	73	2¼	69¼	52
72	86½	2½	82½	60
84	99¾	2¾	95½	64
96	113¼	3	108½	68

Drilling templates are in multiples of four so that fittings may be made to face in any quarter, and bolt hole straddles the center line. All 25-lb. cast iron standard flanges have plane faces.

Templates for Drilling 25-Lb.
Cast Iron Flanged Fittings

Pipe Diameter, Inches	Bolt Diameter, Inches†	Length of Bolts, Inches	Diameters of Ring Gasket, Inches
4	5/8	2¼	4x 6⅞
5	5/8	2¼	5x 7⅞
6	5/8	2¼	6x 8⅞
8	5/8	2¼	8x 11⅛
10	5/8	2½	10x 13⅝
12	5/8	2¾	12x 16⅜
14	3/4	3¼	14x 18
16	3/4	3¼	16x 20½
18	3/4	3½	18x 22
20	3/4	3½	20x 24¼
24	3/4	3¾	24x 28¾
30	7/8	4¼	30x 35⅛
36	7/8	5	36x 41⅞
42	1	5¼	42x 48½
48	1	5½	48x 55
54	1	5¾	54x 61¾
60	1⅛	6	60x 68⅛
72	1⅛	6¼	72x 81⅜
84	1¼	7¼	84x 94¼
96	1¼	7¾	96x107¼

† Bolt hole diameter ⅛ inch larger than bolt throughout

Bolt holes on cast iron flanged fittings are not spot faced for ordinary service.

Templates for Drilling 125-Lb.
Cast Iron Flanges

Pipe Diameter, Inches	Flange Diameter, Inches	Flange Thickness, Inches	Bolt Circle Diameter, Inches	Number of Bolts
1	4¼	⁷⁄₁₆	3⅛	4
1¼	4⅝	½	3½	4
1½	5	⁹⁄₁₆	3⅞	4
2	6	⅝	4¾	4
2½	7	¹¹⁄₁₆	5½	4
3	7½	¾	6	4
3½	8½	¹³⁄₁₆	7	8
4	9	¹⁵⁄₁₆	7½	8
5	10	¹⁵⁄₁₆	8½	8
6	11	1	9½	8
8	13½	1⅛	11¾	8
10	16	1³⁄₁₆	14¼	12
12	19	1¼	17	12
14 OD	21	1⅜	18¾	12
16 OD	23½	1⁷⁄₁₆	21¼	16
18 OD	25	1⁹⁄₁₆	22¾	16
20 OD	27½	1¹¹⁄₁₆	25	20
24 OD	32	1⅞	29½	20
30 OD	38¾	2⅛	36	28
36 OD	46	2⅜	42¾	32
42 OD	53	2⅝	49½	36
48 OD	59½	2¾	56	44
54 OD	66¼	3	62¾	44
60 OD	73	3⅛	69¼	52
72 OD	86½	3½	82½	60
84 OD	99¾	3⅞	95½	64
96 OD	113¼	4¼	108½	68

These cast iron flanges and flanged fittings are plane faced. Bolt holes shall be in multiples of four so that fittings may be made to face in any quarter. The bolt holes shall straddle the center line.

Templates for Drilling 125-Lb. Cast Iron Flanges

Pipe Diameter, Inches	Bolt Diameter, Inches†	Length of Bolts, Inches	Diameters of Ring Gasket, Inches
1	$\frac{1}{2}$	$1\frac{3}{4}$	1 x $2\frac{5}{8}$
$1\frac{1}{4}$	$\frac{1}{2}$	2	$1\frac{1}{4}$x 3
$1\frac{1}{2}$	$\frac{1}{2}$	2	$1\frac{1}{2}$x $3\frac{3}{8}$
2	$\frac{5}{8}$	$2\frac{1}{4}$	2 x $4\frac{1}{8}$
$2\frac{1}{2}$	$\frac{5}{8}$	$2\frac{1}{2}$	$2\frac{1}{2}$x $4\frac{7}{8}$
3	$\frac{5}{8}$	$2\frac{1}{2}$	3 x $5\frac{3}{8}$
$3\frac{1}{2}$	$\frac{5}{8}$	$2\frac{3}{4}$	$3\frac{1}{2}$x $6\frac{3}{8}$
4	$\frac{5}{8}$	3	4 x $6\frac{7}{8}$
5	$\frac{3}{4}$	3	5 x $7\frac{3}{4}$
6	$\frac{3}{4}$	$3\frac{1}{4}$	6 x $8\frac{3}{4}$
8	$\frac{3}{4}$	$3\frac{1}{2}$	8 x 11
10	$\frac{7}{8}$	$3\frac{3}{4}$	10 x $13\frac{3}{8}$
12	$\frac{7}{8}$	$3\frac{3}{4}$	12 x $16\frac{1}{8}$
14 OD	1	$4\frac{1}{4}$	14 x $17\frac{3}{4}$
16 OD	1	$4\frac{1}{2}$	16 x $20\frac{1}{4}$
18 OD	$1\frac{1}{8}$	$4\frac{3}{4}$	18 x $21\frac{5}{8}$
20 OD	$1\frac{1}{8}$	5	20 x $23\frac{7}{8}$
24 OD	$1\frac{1}{4}$	$5\frac{1}{2}$	24 x $28\frac{1}{4}$
30 OD	$1\frac{1}{4}$	$6\frac{1}{4}$	30 x $34\frac{3}{4}$
36 OD	$1\frac{1}{2}$	7	36 x $41\frac{1}{4}$
42 OD	$1\frac{1}{2}$	$7\frac{1}{2}$	42 x 48
48 OD	$1\frac{1}{2}$	$7\frac{3}{4}$	48 x $54\frac{1}{2}$
54 OD	$1\frac{3}{4}$	$8\frac{1}{2}$	54 x 61
60 OD	$1\frac{3}{4}$	$8\frac{3}{4}$	60 x $67\frac{1}{2}$
72 OD	$1\frac{3}{4}$	$9\frac{1}{2}$	72 x $80\frac{3}{4}$
84 OD	2	$10\frac{1}{2}$	84 x $93\frac{1}{2}$
96 OD	$2\frac{1}{4}$	$11\frac{1}{2}$	96 x$106\frac{1}{4}$

† Bolt hole diameter $\frac{1}{8}$ inch larger than bolt except in 54 inch pipe and larger where hole is $\frac{1}{4}$ inch larger than bolt.

Templates for Drilling 250-Lb.
Cast Iron Flanges

Pipe Diameter, Inches	Flange Diameter, Inches	Flange Thickness, Inches	Bolt Circle Diameter, Inches	Number of Bolts
1	$4\frac{7}{8}$	$\frac{11}{16}$	$3\frac{1}{2}$	4
$1\frac{1}{4}$	$5\frac{1}{4}$	$\frac{3}{4}$	$3\frac{7}{8}$	4
$1\frac{1}{2}$	$6\frac{1}{8}$	$\frac{13}{16}$	$4\frac{1}{2}$	4
2	$6\frac{1}{2}$	$\frac{7}{8}$	5	8
$2\frac{1}{2}$	$7\frac{1}{2}$	1	$5\frac{7}{8}$	8
3	$8\frac{1}{4}$	$1\frac{1}{8}$	$6\frac{5}{8}$	8
$3\frac{1}{2}$	9	$1\frac{3}{16}$	$7\frac{1}{4}$	8
4	10	$1\frac{1}{4}$	$7\frac{7}{8}$	8
5	11	$1\frac{3}{8}$	$9\frac{1}{4}$	8
6	$12\frac{1}{2}$	$1\frac{7}{16}$	$10\frac{5}{8}$	12
8	15	$1\frac{5}{8}$	13	12
10	$17\frac{1}{2}$	$1\frac{7}{8}$	$15\frac{1}{4}$	16
12	$20\frac{1}{2}$	2	$17\frac{3}{4}$	16
14 OD	23	$2\frac{1}{8}$	$20\frac{1}{4}$	20
16 OD	$25\frac{1}{2}$	$2\frac{1}{4}$	$22\frac{1}{2}$	20
18 OD	28	$2\frac{3}{8}$	$24\frac{3}{4}$	24
20 OD	$30\frac{1}{2}$	$2\frac{1}{2}$	27	24
24 OD	36	$2\frac{3}{4}$	32	24
30 OD	43	3	$39\frac{1}{4}$	28
36 OD	50	$3\frac{3}{8}$	46	32
42 OD	57	$3\frac{11}{16}$	$52\frac{3}{4}$	36
48 OD	65	4	$60\frac{3}{4}$	40

Templates for Drilling 250-Lb.
Cast Iron Flanges

Pipe Diameter, Inches	Bolt Diameter, Inches†	Length of Bolts, Inches	Diameters of Ring Gaskets, Inches
1	¾	2½	1 x 2⅞
1¼	¾	2½	1¼x 3¼
1½	⅞	2¾	1½x 3¾
2	¾	2¾	2 x 4⅜
2½	⅞	3¼	2½x 5⅛
3	⅞	3½	3 x 5⅞
3½	⅞	3½	3½x 6½
4	⅞	3¾	4 x 7⅛
5	⅞	4	5 x 8½
6	⅞	4	6 x 9⅞
8	1	4½	8 x12⅛
10	1⅛	5¼	10 x14¼
12	1¼	5½	12 x16⅝
14 OD	1¼	6	13¼x19⅛
16 OD	1⅜	6¼	15¼x21¼
18 OD	1⅜	6½	17 x23½
20 OD	1⅜	6¾	19 x25¾
24 OD	1¹¹⁄₁₆	7¾	23 x30½
30 OD	2	8½	29 x37½
36 OD	2¼	9½	34½x44
42 OD	2¼	10¼	40¼x50¾
48 OD	2¼	10¾	46 x58¾

† Bolt hole diameter ⅛ inch larger than bolt except in 24-inch pipe where hole is ³⁄₁₆ larger than bolt and in 30-inch and larger where hole is ¼ inch larger than bolt.

Laying-In Dimensions of
25-Lb. Cast Iron Flanged Elbows

90° Elbow

90° Long
radius elbow

90° Reducing
elbow

45° Elbow

Pipe Diameter, Inches	Dimension A, Inches	Dimension B, Inches	Dimension C, Inches
4	6½	9	4
5	7½	10¼	4½
6	8	11½	5
8	9	14	5½
10	11	16½	6½
12	12	19	7½
14	14	21½	7½
16	15	24	8
18	16½	26½	8½
20	18	29	9½
24	22	34	11
30	25	41½	15
36	28	49	18
42	31	56½	21
48	34	64	24
54	39	71½	27
60	44	79	30
72	53	94	36

Laying-In Dimensions of
25-Lb. Cast Iron Flanged Tees and Crosses

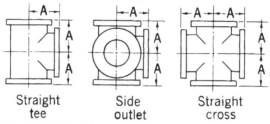

Straight   Side   Straight
tee   outlet   cross

Pipe Diameter, Inches	Center-to-Face, A, Inches	Face-to-Face, 2A, Inches	Flange Diameter, Inches
4	6½	13	9
5	7½	15	10
6	8	16	11
8	9	18	13½
10	11	22	16
12	12	24	19
14	14	28	21
16	15	30	23½
18	16½	33	25
20	18	36	27½
24	22	44	32
30	25	50	38¾
36	28	56	46
42	31	62	53
48	34	68	59½
54	39	78	66¼
60	44	88	73
72	53	106	86½

Laying-In Dimensions of
25-Lb. Flanged Reducing Tees and Crosses

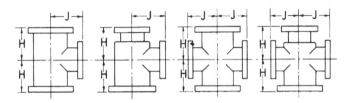

All reducing fittings sizes 16 inches and smaller have same center-to-face dimensions as straight size fittings.

Pipe Diameter, Inches	Size of Outlet and Smaller, Inches	Center-to-Face, H, Inches	Face-to-Face, 2H, Inches	Center-to-Face, J, Inches
18	12	13	26	15½
20	14	14	28	17
24	16	15	30	19
30	20	18	36	23
36	24	20	40	26
42	24	23	46	30
48	30	26	52	34
54	36	29	58	37
60	40	33	66	41
72	48	40	80	48

Note: Fittings reducing on the run only carry same dimensions center-to-face and face-to-face as straight size fittings corresponding to size of the larger opening. Tees increasing on outlet, known as Bull Head Tees, have same center-to-face and face-to-face dimensions as a straight fitting of the size of the outlet. For example: a 12 x 12 x 18 inch tee is governed by dimensions of the 18 inch straight tee; namely, 16½ inches center-to-face of all openings and 33 inches face-to-face.

Dimensions of 125-Lb. Cast Iron
Screwed Companion and Blind Flanges

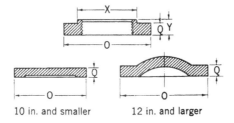

10 in. and smaller   12 in. and larger

Pipe Diameter, I, Inches	Flange Diameter, O, Inches	Flange Thickness, Q, Inches	Hub Diameter, X, Inches	Overall Length, Y, Inches
1	$4\frac{1}{4}$	$\frac{7}{16}$	$1\frac{15}{16}$	$\frac{11}{16}$
$1\frac{1}{4}$	$4\frac{5}{8}$	$\frac{1}{2}$	$2\frac{5}{16}$	$\frac{13}{16}$
$1\frac{1}{2}$	5	$\frac{9}{16}$	$2\frac{9}{16}$	$\frac{7}{8}$
2	6	$\frac{5}{8}$	$3\frac{1}{16}$	1
$2\frac{1}{2}$	7	$\frac{11}{16}$	$3\frac{9}{16}$	$1\frac{1}{8}$
3	$7\frac{1}{2}$	$\frac{3}{4}$	$4\frac{1}{4}$	$1\frac{3}{16}$
$3\frac{1}{2}$	$8\frac{1}{2}$	$\frac{13}{16}$	$4\frac{13}{16}$	$1\frac{1}{4}$
4	9	$\frac{15}{16}$	$5\frac{5}{16}$	$1\frac{5}{16}$
5	10	$\frac{15}{16}$	$6\frac{7}{16}$	$1\frac{7}{16}$
6	11	1	$7\frac{9}{16}$	$1\frac{9}{16}$
8	$13\frac{1}{2}$	$1\frac{1}{8}$	$9\frac{11}{16}$	$1\frac{3}{4}$
10	16	$1\frac{3}{16}$	$11\frac{15}{16}$	$1\frac{15}{16}$
12	19	$1\frac{1}{4}$	$14\frac{1}{16}$	$2\frac{3}{16}$
14 OD	21	$1\frac{3}{8}$	$15\frac{3}{8}$	$2\frac{1}{4}$
16 OD	$23\frac{1}{2}$	$1\frac{7}{16}$	$17\frac{1}{2}$	$2\frac{1}{2}$
18 OD	25	$1\frac{9}{16}$	$19\frac{5}{8}$	$2\frac{11}{16}$
20 OD	$27\frac{1}{2}$	$1\frac{11}{16}$	$21\frac{3}{4}$	$2\frac{7}{8}$
24 OD	32	$1\frac{7}{8}$	26	$3\frac{1}{4}$
30 OD	$38\frac{3}{4}$	$2\frac{1}{8}$	—	—
36 OD	46	$2\frac{3}{8}$	—	—
42 OD	53	$2\frac{5}{8}$	—	—
48 OD	$59\frac{1}{2}$	$2\frac{3}{4}$	—	—

Laying Lengths of
125-Lb. Cast Iron Flanged Elbows and Crosses

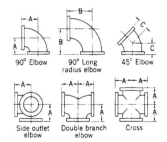

90° Elbow   90° Long   45° Elbow
             radius elbow

Side outlet   Double branch   Cross
   elbow   elbow

Pipe Diameter, Inches	Dimension A, Inches	Dimension B, Inches	Dimension C, Inches
1	$3\frac{1}{2}$	5	$1\frac{3}{4}$
$1\frac{1}{4}$	$3\frac{3}{4}$	$5\frac{1}{2}$	2
$1\frac{1}{2}$	4	6	$2\frac{1}{4}$
2	$4\frac{1}{2}$	$6\frac{1}{2}$	$2\frac{1}{2}$
$2\frac{1}{2}$	5	7	3
3	$5\frac{1}{2}$	$7\frac{3}{4}$	3
$3\frac{1}{2}$	6	$8\frac{1}{2}$	$3\frac{1}{2}$
4	$6\frac{1}{2}$	9	4
5	$7\frac{1}{2}$	$10\frac{1}{4}$	$4\frac{1}{2}$
6	8	$11\frac{1}{2}$	5
8	9	14	$5\frac{1}{2}$
10	11	$16\frac{1}{2}$	$6\frac{1}{2}$
12	12	19	$7\frac{1}{2}$
14	14	$21\frac{1}{2}$	$7\frac{1}{2}$
16	15	24	8
18	$16\frac{1}{2}$	$26\frac{1}{2}$	$8\frac{1}{2}$
20	18	29	$9\frac{1}{2}$

Laying Lengths of 125-Lb. Cast Iron Flanged Miscellaneous Fittings

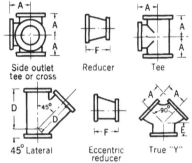

Side outlet tee or cross   Reducer   Tee

45° Lateral   Eccentric reducer   True "Y"

Pipe Diameter, Inches	Dimension A, Inches	Dimension D, Inches	Dimension E, Inches	Dimension F, Inches
1	$3\frac{1}{2}$	$5\frac{3}{4}$	$1\frac{3}{4}$	—
$1\frac{1}{4}$	$3\frac{3}{4}$	$6\frac{1}{4}$	$1\frac{3}{4}$	—
$1\frac{1}{2}$	4	7	2	—
2	$4\frac{1}{2}$	8	$2\frac{1}{2}$	5
$2\frac{1}{2}$	5	$9\frac{1}{2}$	$2\frac{1}{2}$	$5\frac{1}{2}$
3	$5\frac{1}{2}$	10	3	6
$3\frac{1}{2}$	6	$11\frac{1}{2}$	3	$6\frac{1}{2}$
4	$6\frac{1}{2}$	12	3	7
5	$7\frac{1}{2}$	$13\frac{1}{2}$	$3\frac{1}{2}$	8
6	8	$14\frac{1}{2}$	$3\frac{1}{2}$	9
8	9	$17\frac{1}{2}$	$4\frac{1}{2}$	11
10	11	$20\frac{1}{2}$	5	12
12	12	$24\frac{1}{2}$	$5\frac{1}{2}$	14
14	14	27	6	16
16	15	30	$6\frac{1}{2}$	18
18	$16\frac{1}{2}$	32	7	19
20	18	35	8	20
24	22	$40\frac{1}{2}$	9	24
30	25	49	10	30

Laying Lengths of
125-Lb. Cast Iron
Flanged Short Body Reducing Tees and Crosses

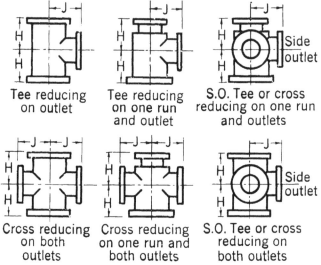

Tee reducing on outlet   Tee reducing on one run and outlet   S.O. Tee or cross reducing on one run and outlets

Cross reducing on both outlets   Cross reducing on one run and both outlets   S.O. Tee or cross reducing on both outlets

All reducing tees and crosses, sizes 16-inch and smaller, have same center-to-face dimensions as straight size fittings, corresponding to the size of the largest opening.

Pipe Diameter, Inches	Outlet Size and Smaller, Inches	Dimension H, Inches	Dimension J, Inches
18	12	13	15½
20	14	14	17
24	16	15	19
30	20	18	23
36	24	20	26
42	24	23	30
48	30	26	34

Laying Lengths of
125-Lb. Cast Iron
Flanged Short Body Reducing Laterals

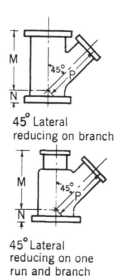

45° Lateral
reducing on branch

45° Lateral
reducing on one
run and branch

All reducing laterals, sizes 16-inch and smaller, have same center-to-face dimensions as straight size fittings corresponding to size of the largest opening.

Pipe Diameter, Inches	Size of Branch and Smaller, Inches	Dimension M, Inches	Dimension N, Inches	Dimension P, Inches
18	8	25	1	27½
20	10	27	1	29½
24	12	31½	½	34½
30	14	39	0	42

Laying Lengths of 125-Lb. Cast Iron Flanged Base Elbows and Tees

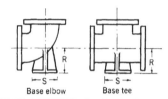

Base elbow   Base tee

Pipe Diameter, Inches	Dimension R, Inches	Dia. or Width of Base, S, Inches	Size of Supporting Pipe, Inches
1	$3\frac{1}{2}$	$3\frac{1}{2}$	$\frac{3}{4}$
$1\frac{1}{4}$	$3\frac{5}{8}$	$3\frac{1}{2}$	$\frac{3}{4}$
$1\frac{1}{2}$	$3\frac{3}{4}$	$4\frac{1}{4}$	1
2	$4\frac{1}{8}$	$4\frac{5}{8}$	$1\frac{1}{4}$
$2\frac{1}{2}$	$4\frac{1}{2}$	$4\frac{5}{8}$	$1\frac{1}{4}$
3	$4\frac{7}{8}$	5	$1\frac{1}{2}$
4	$5\frac{1}{2}$	6	2
5	$6\frac{1}{4}$	7	$2\frac{1}{2}$
6	7	7	$2\frac{1}{2}$
8	$8\frac{3}{8}$	9	4
10	$9\frac{3}{4}$	9	4
12	$11\frac{1}{4}$	11	6
14	$12\frac{1}{2}$	11	6
16	$13\frac{3}{4}$	11	6
18	15	$13\frac{1}{2}$	8
20	16	$13\frac{1}{2}$	8
24	$18\frac{1}{2}$	$13\frac{1}{2}$	8

Laying Lengths of
125-Lb. Cast Iron
Flanged Anchorage Bases for Tees

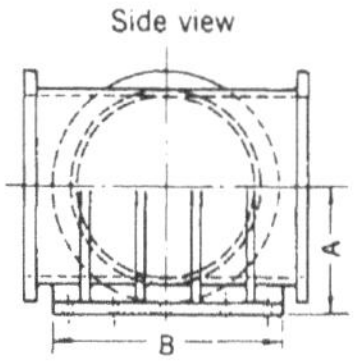

Side view

Pipe Diameter, Inches	Center-to-Base, A, Inches	Width and Length of Base, B, Inches
2½	4½	7
3	4⅞	7½
3½	5¼	8½
4	5½	9
5	6¼	10
6	7	11
8	8⅜	13½
10	9¾	16
12	11¼	19
14 OD	12½	21
16 OD	13¾	23½
18 OD	15	25
20 OD	16	27½
24 OD	18½	32
30 OD	22	38¾
36 OD	25½	46
42 OD	29¼	53

Dimensions of 250-Lb. Cast Iron
Screwed Companion and Blind Flanges

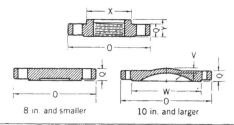

8 in. and smaller　　　　　　　　10 in. and larger

Pipe Diameter, Inches	Flange Diameter, O, Inches	Flange Thickness, Q, Inches	Hub Diameter, X, Inches	Overall Length, Y, Inches
1	4⅞	¹¹⁄₁₆	2¹⁄₁₆	⅞
1¼	5¼	¾	2½	1
1½	6⅛	¹³⁄₁₆	2¾	1⅛
2	6½	⅞	3⁵⁄₁₆	1¼
2½	7½	1	3¹⁵⁄₁₆	1⁷⁄₁₆
3	8¼	1⅛	4⅝	1⁹⁄₁₆
3½	9	1³⁄₁₆	5¼	1⅝
4	10	1¼	5¾	1¾
5	11	1⅜	7	1⅞
6	12½	1⁷⁄₁₆	8⅛	1¹⁵⁄₁₆
8	15	1⅝	10¼	2³⁄₁₆
10	17½	1⅞	12⅝	2⅜
12	20½	2	14¾	2⁹⁄₁₆
14 OD	23	2⅛	16¼	2¹¹⁄₁₆
16 OD	25½	2¼	18⅜	2⅞
18 OD	28	2⅜	—	—
20 OD	30½	2½	—	—
24 OD	36	2¾	—	—

Laying Lengths of 250-Lb. Cast Iron Flanged Elbows, Tees and Reducers

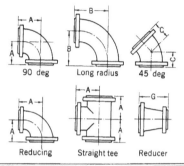

90 deg   Long radius   45 deg

Reducing   Straight tee   Reducer

Pipe Diameter, Inches	Dimension A, Inches	Dimension B, Inches	Dimension C, Inches	Dimension G, Inches
2	5	$6\frac{1}{2}$	3	5
$2\frac{1}{2}$	$5\frac{1}{2}$	7	$3\frac{1}{2}$	$5\frac{1}{2}$
3	6	$7\frac{3}{4}$	$3\frac{1}{2}$	6
$3\frac{1}{2}$	$6\frac{1}{2}$	$8\frac{1}{2}$	4	$6\frac{1}{2}$
4	7	9	$4\frac{1}{2}$	7
5	8	$10\frac{1}{4}$	5	8
6	$8\frac{1}{2}$	$11\frac{1}{2}$	$5\frac{1}{2}$	9
8	10	14	6	11
10	$11\frac{1}{2}$	$16\frac{1}{2}$	7	12
12	13	19	8	14
14 OD	15	$21\frac{1}{2}$	$8\frac{1}{2}$	16
16 OD	$16\frac{1}{2}$	24	$9\frac{1}{2}$	18
18 OD	18	$26\frac{1}{2}$	10	19
20 OD	$19\frac{1}{2}$	29	$10\frac{1}{2}$	20
24 OD	$22\frac{1}{2}$	34	12	24

Laying Lengths of
250-Lb. Cast Iron
Flanged Base Elbows and Tees

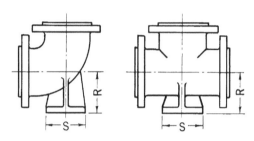

Pipe Diameter, Inches	Center-to-Base, R, Inches	Diameter of Round Base, S, Inches	Size of Supporting Pipe, Inches
2	4½	5¼	1¼
2½	4¾	5¼	1¼
3	5¼	6⅛	1½
3½	5⅝	6⅛	1½
4	6	6½	2
5	6¾	7½	2½
6	7½	7½	2½
8	9	10	4
10	10½	10	4
12	12	12½	6
14 OD	13½	12½	6
16 OD	14¾	12½	6
18 OD	16¼	15	8
20 OD	17⅞	15	8
24 OD	20¾	17½	10

Laying Lengths of 150-, 300- and 400-Lb. Steel Welding Neck Flanges

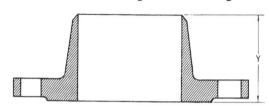

Nominal Diameter, Inches	Length of Y, Inches		
	150-Lb.	300-Lb.	400-Lb.
$\frac{1}{2}$	$1\frac{7}{8}$	$2\frac{1}{16}$	$2\frac{1}{16}$
$\frac{3}{4}$	$2\frac{1}{16}$	$2\frac{1}{4}$	$2\frac{1}{4}$
1	$2\frac{3}{16}$	$2\frac{7}{16}$	$2\frac{7}{16}$
$1\frac{1}{4}$	$2\frac{1}{4}$	$2\frac{9}{16}$	$2\frac{5}{8}$
$1\frac{1}{2}$	$2\frac{7}{16}$	$2\frac{11}{16}$	$2\frac{3}{4}$
2	$2\frac{1}{2}$	$2\frac{3}{4}$	$2\frac{7}{8}$
$2\frac{1}{2}$	$2\frac{3}{4}$	3	$3\frac{1}{8}$
3	$2\frac{3}{4}$	$3\frac{1}{8}$	$3\frac{1}{4}$
$3\frac{1}{2}$	$2\frac{13}{16}$	$3\frac{3}{16}$	$3\frac{3}{8}$
4	3	$3\frac{3}{8}$	$3\frac{1}{2}$
5	$3\frac{1}{2}$	$3\frac{7}{8}$	4
6	$3\frac{1}{2}$	$3\frac{7}{8}$	$4\frac{1}{16}$
8	4	$4\frac{3}{8}$	$4\frac{5}{8}$
10	4	$4\frac{5}{8}$	$4\frac{7}{8}$
12	$4\frac{1}{2}$	$5\frac{1}{8}$	$5\frac{3}{8}$
14 OD	5	$5\frac{5}{8}$	$5\frac{7}{8}$
16 OD	5	$5\frac{3}{4}$	6
18 OD	$5\frac{1}{2}$	$6\frac{1}{4}$	$6\frac{1}{2}$
20 OD	$5\frac{11}{16}$	$6\frac{3}{8}$	$6\frac{5}{8}$
24 OD	6	$6\frac{5}{8}$	$6\frac{7}{8}$

Laying Lengths of
600-, 900-, 1500- and 2500-Lb.
Steel Welding Neck Flanges

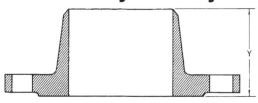

Nominal Diameter, Inches	Length of Y, Inches			
	600-Lb.	900-Lb.	1500-Lb.	2500-Lb.
$\frac{1}{2}$	$2\frac{1}{16}$	$2\frac{3}{8}$	$2\frac{3}{8}$	$2\frac{7}{8}$
$\frac{3}{4}$	$2\frac{1}{4}$	$2\frac{3}{4}$	$2\frac{3}{4}$	$3\frac{1}{8}$
1	$2\frac{7}{16}$	$2\frac{7}{8}$	$2\frac{7}{8}$	$3\frac{1}{2}$
$1\frac{1}{4}$	$2\frac{5}{8}$	$2\frac{7}{8}$	$2\frac{7}{8}$	$3\frac{3}{4}$
$1\frac{1}{2}$	$2\frac{3}{4}$	$3\frac{1}{4}$	$3\frac{1}{4}$	$4\frac{3}{8}$
2	$2\frac{7}{8}$	4	4	5
$2\frac{1}{2}$	$3\frac{1}{8}$	$4\frac{1}{8}$	$4\frac{1}{8}$	$5\frac{5}{8}$
3	$3\frac{1}{4}$	4	$4\frac{5}{8}$	$6\frac{5}{8}$
$3\frac{1}{2}$	$3\frac{3}{8}$	—	—	—
4	4	$4\frac{1}{2}$	$4\frac{7}{8}$	$7\frac{1}{2}$
5	$4\frac{1}{2}$	5	$6\frac{1}{8}$	9
6	$4\frac{5}{8}$	$5\frac{1}{2}$	$6\frac{3}{4}$	$10\frac{3}{4}$
8	$5\frac{1}{4}$	$6\frac{3}{8}$	$8\frac{3}{8}$	$12\frac{1}{2}$
10	6	$7\frac{1}{4}$	10	$16\frac{1}{2}$
12	$6\frac{1}{8}$	$7\frac{7}{8}$	$11\frac{1}{8}$	$18\frac{1}{4}$
14 OD	$6\frac{1}{2}$	$8\frac{3}{8}$	$11\frac{3}{4}$	—
16 OD	7	$8\frac{1}{2}$	$12\frac{1}{4}$	—
18 OD	$7\frac{1}{4}$	9	$12\frac{7}{8}$	—
20 OD	$7\frac{1}{2}$	$9\frac{3}{4}$	14	—
24 OD	8	$11\frac{1}{2}$	16	—

Laying Lengths of
150-Lb. Steel Flanges

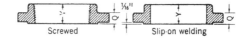

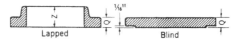

Pipe Diameter, Inches	Flange Thickness, Q, Inches	Screwed Flange, Y, Inches	Lapped Flange, Z, Inches
$\frac{1}{2}$	$\frac{7}{16}$	$\frac{5}{8}$	$\frac{5}{8}$
$\frac{3}{4}$	$\frac{1}{2}$	$\frac{5}{8}$	$\frac{5}{8}$
1	$\frac{9}{16}$	$\frac{11}{16}$	$\frac{11}{16}$
$1\frac{1}{4}$	$\frac{5}{8}$	$\frac{13}{16}$	$\frac{13}{16}$
$1\frac{1}{2}$	$\frac{11}{16}$	$\frac{7}{8}$	$\frac{7}{8}$
2	$\frac{3}{4}$	1	1
$2\frac{1}{2}$	$\frac{7}{8}$	$1\frac{1}{8}$	$1\frac{1}{8}$
3	$\frac{15}{16}$	$1\frac{3}{16}$	$1\frac{3}{16}$
$3\frac{1}{2}$	$\frac{15}{16}$	$1\frac{1}{4}$	$1\frac{1}{4}$
4	$\frac{15}{16}$	$1\frac{5}{16}$	$1\frac{5}{16}$
5	$\frac{15}{16}$	$1\frac{7}{16}$	$1\frac{7}{16}$
6	1	$1\frac{9}{16}$	$1\frac{9}{16}$
8	$1\frac{1}{8}$	$1\frac{3}{4}$	$1\frac{3}{4}$
10	$1\frac{3}{16}$	$1\frac{15}{16}$	$1\frac{15}{16}$
12	$1\frac{1}{4}$	$2\frac{3}{16}$	$2\frac{3}{16}$
14 OD	$1\frac{3}{8}$	$2\frac{1}{4}$	$3\frac{1}{8}$
16 OD	$1\frac{7}{16}$	$2\frac{1}{2}$	$3\frac{7}{16}$
18 OD	$1\frac{9}{16}$	$2\frac{11}{16}$	$3\frac{13}{16}$
20 OD	$1\frac{11}{16}$	$2\frac{7}{8}$	$4\frac{1}{16}$
24 OD	$1\frac{7}{8}$	$3\frac{1}{4}$	$4\frac{3}{8}$

Note: The $\frac{1}{16}$ inch raised face is included in Q and Y.

Laying Lengths of 150-Lb. Steel
Flanged Elbows, Tees and Crosses
($\frac{1}{16}$ Inch Raised Face Type)

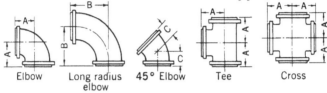

| Elbow | Long radius elbow | 45° Elbow | Tee | Cross |

Pipe Diameter, Inches	Dimension A, Inches	Dimension B, Inches	Dimension C, Inches
1	$3\frac{1}{2}$	5	$1\frac{3}{4}$
$1\frac{1}{4}$	$3\frac{3}{4}$	$5\frac{1}{2}$	2
$1\frac{1}{2}$	4	6	$2\frac{1}{4}$
2	$4\frac{1}{2}$	$6\frac{1}{2}$	$2\frac{1}{2}$
$2\frac{1}{2}$	5	7	3
3	$5\frac{1}{2}$	$7\frac{3}{4}$	3
$3\frac{1}{2}$	6	$8\frac{1}{2}$	$3\frac{1}{2}$
4	$6\frac{1}{2}$	9	4
5	$7\frac{1}{2}$	$10\frac{1}{4}$	$4\frac{1}{2}$
6	8	$11\frac{1}{2}$	5
8	9	14	$5\frac{1}{2}$
10	11	$16\frac{1}{2}$	$6\frac{1}{2}$
12	12	19	$7\frac{1}{2}$
14 OD	14	$21\frac{1}{2}$	$7\frac{1}{2}$
16 OD	15	24	8
18 OD	$16\frac{1}{2}$	$26\frac{1}{2}$	$8\frac{1}{2}$
20 OD	18	29	$9\frac{1}{2}$
24 OD	22	34	11

Laying Lengths of 150-Lb. Steel Flanged Laterals and Reducers

(1/16 Inch Raised Face Type)

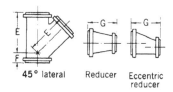

45° lateral   Reducer   Eccentric reducer

Pipe Diameter, Inches	Dimension E, Inches	Dimension F, Inches	Dimension G, Inches
1	5¾	1¾	—
1¼	6¼	1¾	—
1½	7	2	—
2	8	2½	5
2½	9½	2½	5½
3	10	3	6
3½	11½	3	6½
4	12	3	7
5	13½	3½	8
6	14½	3½	9
8	17½	4½	11
10	20½	5	12
12	24½	5½	14
14 OD	27	6	16
16 OD	30	6½	18
18 OD	32	7	19
20 OD	35	8	20
24 OD	40½	9	24

Laying Lengths of 150-Lb. Steel Flanged Elbows, Tees and Crosses
(Ring Joint Type)

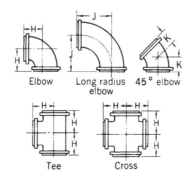

Pipe Diameter, Inches	Dimension H, Inches	Dimension J, Inches	Dimension K, Inches
1	$3\frac{23}{32}$	$5\frac{7}{32}$	$1\frac{31}{32}$
$1\frac{1}{4}$	$3\frac{31}{32}$	$5\frac{23}{32}$	$2\frac{7}{32}$
$1\frac{1}{2}$	$4\frac{7}{32}$	$6\frac{7}{32}$	$2\frac{15}{32}$
2	$4\frac{3}{4}$	$6\frac{3}{4}$	$2\frac{3}{4}$
$2\frac{1}{2}$	$5\frac{1}{4}$	$7\frac{1}{4}$	$3\frac{1}{4}$
3	$5\frac{3}{4}$	8	$3\frac{1}{4}$
$3\frac{1}{2}$	$6\frac{1}{4}$	$8\frac{3}{4}$	$3\frac{3}{4}$
4	$6\frac{3}{4}$	$9\frac{1}{4}$	$4\frac{1}{4}$
5	$7\frac{3}{4}$	$10\frac{1}{2}$	$4\frac{3}{4}$
6	$8\frac{1}{4}$	$11\frac{3}{4}$	$5\frac{1}{4}$
8	$9\frac{1}{4}$	$14\frac{1}{4}$	$5\frac{3}{4}$
10	$11\frac{1}{4}$	$16\frac{3}{4}$	$6\frac{3}{4}$
12	$12\frac{1}{4}$	$19\frac{1}{4}$	$7\frac{3}{4}$
14 OD	$14\frac{1}{4}$	$21\frac{3}{4}$	$7\frac{3}{4}$
16 OD	$15\frac{1}{4}$	$24\frac{1}{4}$	$8\frac{1}{4}$
18 OD	$16\frac{3}{4}$	$26\frac{3}{4}$	$8\frac{3}{4}$
20 OD	$18\frac{1}{4}$	$29\frac{1}{4}$	$9\frac{3}{4}$
24 OD	$22\frac{1}{4}$	$34\frac{1}{4}$	$11\frac{1}{4}$

Laying Lengths of 150-Lb. Steel Flanged Laterals and Reducers
(Ring Joint Type)

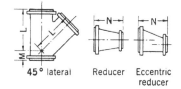

45° lateral   Reducer   Eccentric
                          reducer

Pipe Diameter, Inches	Dimension L, Inches	Dimension M, Inches	Dimension N, Inches
1	$5\frac{31}{32}$	$1\frac{31}{32}$	—
$1\frac{1}{4}$	$6\frac{15}{32}$	$1\frac{31}{32}$	—
$1\frac{1}{2}$	$7\frac{7}{32}$	$2\frac{7}{32}$	—
2	$8\frac{1}{4}$	$2\frac{3}{4}$	$5\frac{1}{2}$
$2\frac{1}{2}$	$9\frac{3}{4}$	$2\frac{3}{4}$	6
3	$10\frac{1}{4}$	$3\frac{1}{4}$	$6\frac{1}{2}$
$3\frac{1}{2}$	$11\frac{3}{4}$	$3\frac{1}{4}$	7
4	$12\frac{1}{4}$	$3\frac{1}{4}$	$7\frac{1}{2}$
5	$13\frac{3}{4}$	$3\frac{3}{4}$	$8\frac{1}{2}$
6	$14\frac{3}{4}$	$3\frac{3}{4}$	$9\frac{1}{2}$
8	$17\frac{3}{4}$	$4\frac{3}{4}$	$11\frac{1}{2}$
10	$20\frac{3}{4}$	$5\frac{1}{4}$	$12\frac{1}{2}$
12	$24\frac{3}{4}$	$5\frac{3}{4}$	$14\frac{1}{2}$
14 OD	$27\frac{1}{4}$	$6\frac{1}{4}$	$16\frac{1}{2}$
16 OD	$30\frac{1}{4}$	$6\frac{3}{4}$	$18\frac{1}{2}$
18 OD	$32\frac{1}{4}$	$7\frac{1}{4}$	$19\frac{1}{2}$
20 OD	$35\frac{1}{4}$	$8\frac{1}{4}$	$20\frac{1}{2}$
24 OD	$40\frac{3}{4}$	$9\frac{1}{4}$	$24\frac{1}{2}$

Laying Lengths of
150-Lb. Steel
Flanged Base Elbows and Tees

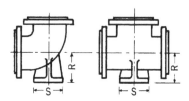

Pipe Diameter, Inches	Center-to-Base, R, Inches	Diameter of Round Base, S, Inches
2	$4\frac{1}{8}$	$4\frac{5}{8}$
$2\frac{1}{2}$	$4\frac{1}{2}$	$4\frac{5}{8}$
3	$4\frac{7}{8}$	5
$3\frac{1}{2}$	$5\frac{1}{4}$	5
4	$5\frac{1}{2}$	6
5	$6\frac{1}{4}$	7
6	7	7
8	$8\frac{3}{8}$	9
10	$9\frac{3}{4}$	9
12	$11\frac{1}{4}$	11
14 OD	$12\frac{1}{2}$	11
16 OD	$13\frac{3}{4}$	11
18 OD	15	$13\frac{1}{2}$
20 OD	16	$13\frac{1}{2}$
24 OD	$18\frac{1}{2}$	$13\frac{1}{2}$

Laying Lengths of 300-Lb. Steel Flanges

Screwed   Slip-on-welding

Lapped   Blind

Pipe Diameter, Inches	Flange Thickness Q, Inches	Screwed Flange Y, Inches	Lapped Flange Z, Inches
$\frac{1}{2}$	$\frac{9}{16}$	$\frac{7}{8}$	$\frac{7}{8}$
$\frac{3}{4}$	$\frac{5}{8}$	1	1
1	$\frac{11}{16}$	$1\frac{1}{16}$	$1\frac{1}{16}$
$1\frac{1}{4}$	$\frac{3}{4}$	$1\frac{1}{16}$	$1\frac{1}{16}$
$1\frac{1}{2}$	$\frac{13}{16}$	$1\frac{3}{16}$	$1\frac{3}{16}$
2	$\frac{7}{8}$	$1\frac{5}{16}$	$1\frac{5}{16}$
$2\frac{1}{2}$	1	$1\frac{1}{2}$	$1\frac{1}{2}$
3	$1\frac{1}{8}$	$1\frac{11}{16}$	$1\frac{11}{16}$
$3\frac{1}{2}$	$1\frac{3}{16}$	$1\frac{3}{4}$	$1\frac{3}{4}$
4	$1\frac{1}{4}$	$1\frac{7}{8}$	$1\frac{7}{8}$
5	$1\frac{3}{8}$	2	2
6	$1\frac{7}{16}$	$2\frac{1}{16}$	$2\frac{1}{16}$
8	$1\frac{5}{8}$	$2\frac{7}{16}$	$2\frac{7}{16}$
10	$1\frac{7}{8}$	$2\frac{5}{8}$	$3\frac{3}{4}$
12	2	$2\frac{7}{8}$	4
14 OD	$2\frac{1}{8}$	3	$4\frac{3}{8}$
16 OD	$2\frac{1}{4}$	$3\frac{1}{4}$	$4\frac{3}{4}$
18 OD	$2\frac{3}{8}$	$3\frac{1}{2}$	$5\frac{1}{8}$
20 OD	$2\frac{1}{2}$	$3\frac{3}{4}$	$5\frac{1}{2}$
24 OD	$2\frac{3}{4}$	$4\frac{3}{16}$	6

Laying Lengths of 300-Lb. Steel
Flanged Elbows, Tees and Crosses
(¹⁄₁₆ Inch Raised Face Type)

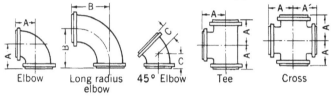

| Elbow | Long radius elbow | 45° Elbow | Tee | Cross |

Pipe Diameter, Inches	Dimension A, Inches	Dimension B, Inches	Dimension C, Inches
1	4	5	2¼
1¼	4¼	5½	2½
1½	4½	6	2¾
2	5	6½	3
2½	5½	7	3½
3	6	7¾	3½
3½	6½	8½	4
4	7	9	4½
5	8	10¼	5
6	8½	11½	5½
8	10	14	6
10	11½	16½	7
12	13	19	8
14 OD	15	21½	8½
16 OD	16½	24	9½
18 OD	18	26½	10
20 OD	19½	29	10½
24 OD	22½	34	12

Laying Lengths of 300-Lb. Steel Flanged Laterals and Reducers
($\frac{1}{16}$ Inch Raised Face Type)

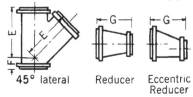

45° lateral   Reducer   Eccentric Reducer

Pipe Diameter, Inches	Dimension E, Inches	Dimension F, Inches	Dimension G, Inches
1	6½	2	4½
1¼	7¼	2¼	4½
1½	8½	2½	4½
2	9	2½	5
2½	10½	2½	5½
3	11	3	6
3½	12½	3	6½
4	13½	3	7
5	15	3½	8
6	17½	4	9
8	20½	5	11
10	24	5½	12
12	27½	6	14
14 OD	31	6½	16
16 OD	34½	7½	18
18 OD	37½	8	19
20 OD	40½	8½	20
24 OD	47½	10	24

Laying Lengths of 300-Lb. Steel Flanged Elbows, Tees and Crosses

(Ring Joint Type)

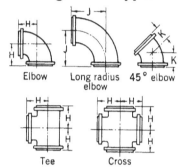

Elbow   Long radius   45° elbow
              elbow

Tee   Cross

Pipe Diameter, Inches	Dimension H, Inches	Dimension J, Inches	Dimension K, Inches
2	5⁵⁄₁₆	6¹³⁄₁₆	3⁵⁄₁₆
2½	5¹³⁄₁₆	7⁵⁄₁₆	3¹³⁄₁₆
3	6⁵⁄₁₆	8¹⁄₁₆	3¹³⁄₁₆
3½	6¹³⁄₁₆	8¹³⁄₁₆	4⁵⁄₁₆
4	7⁵⁄₁₆	9⁵⁄₁₆	4¹³⁄₁₆
5	8⁵⁄₁₆	10⁹⁄₁₆	5⁵⁄₁₆
6	8¹³⁄₁₆	11¹³⁄₁₆	5¹³⁄₁₆
8	10⁵⁄₁₆	14⁵⁄₁₆	6⁵⁄₁₆
10	11¹³⁄₁₆	16¹³⁄₁₆	7⁵⁄₁₆
12	13⁵⁄₁₆	19⁵⁄₁₆	8⁵⁄₁₆
14 OD	15⁵⁄₁₆	21¹³⁄₁₆	8¹³⁄₁₆
16 OD	16¹³⁄₁₆	24⁵⁄₁₆	9¹³⁄₁₆
18 OD	18⁵⁄₁₆	26¹³⁄₁₆	10⁵⁄₁₆
20 OD	19⅞	29⅜	10⅞
24 OD	22¹⁵⁄₁₆	34⁷⁄₁₆	12⁷⁄₁₆

Laying Lengths of 300-Lb. Steel Flanged Laterals and Reducers
(Ring Joint Type)

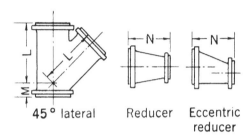

45° lateral   Reducer   Eccentric reducer

Pipe Diameter, Inches	Dimension L, Inches	Dimension M, Inches	Dimension N, Inches
2	9⁵⁄₁₆	2¹³⁄₁₆	5⅝
2½	10¹³⁄₁₆	2¹³⁄₁₆	6⅛
3	11⁵⁄₁₆	3⁵⁄₁₆	6⅝
3½	12¹³⁄₁₆	3⁵⁄₁₆	7⅛
4	13¹³⁄₁₆	3⁵⁄₁₆	7⅝
5	15⁵⁄₁₆	3¹³⁄₁₆	8⅝
6	17¹³⁄₁₆	4⁵⁄₁₆	9⅝
8	20¹³⁄₁₆	5⁵⁄₁₆	11⅝
10	24⁵⁄₁₆	5¹³⁄₁₆	12⅝
12	27¹³⁄₁₆	6⁵⁄₁₆	14⅝
14 OD	31⁵⁄₁₆	6¹³⁄₁₆	16⅝
16 OD	34¹³⁄₁₆	7¹³⁄₁₆	18⅝
18 OD	37¹³⁄₁₆	8⁵⁄₁₆	19⅝
20 OD	40⅞	8⅞	20¾
24 OD	47¹⁵⁄₁₆	10⁷⁄₁₆	24⅞

Laying Lengths of
300-Lb. Steel
Flanged Base Elbows and Tees

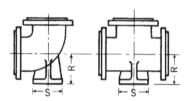

Pipe Diameter, Inches	Center-to-Base, R, Inches	Diameter of Round Base, S, Inches
2	$4\frac{1}{2}$	$5\frac{1}{4}$
$2\frac{1}{2}$	$4\frac{3}{4}$	$5\frac{1}{4}$
3	$5\frac{1}{4}$	$6\frac{1}{8}$
$3\frac{1}{2}$	$5\frac{5}{8}$	$6\frac{1}{8}$
4	6	$6\frac{1}{2}$
5	$6\frac{3}{4}$	$7\frac{1}{2}$
6	$7\frac{1}{2}$	$7\frac{1}{2}$
8	9	10
10	$10\frac{1}{2}$	10
12	12	$12\frac{1}{2}$
14 OD	$13\frac{1}{2}$	$12\frac{1}{2}$
16 OD	$14\frac{3}{4}$	$12\frac{1}{2}$
18 OD	$16\frac{1}{4}$	15
20 OD	$17\frac{7}{8}$	15
24 OD	$20\frac{3}{4}$	$17\frac{1}{2}$

Laying Lengths of
400-Lb. Steel Flanges

Blind

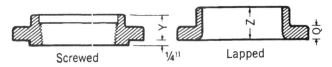

Screwed   Lapped

Pipe Diameter, Inches	Flange Thickness, Q, Inches	Screwed Flange, Y, Inches	Lapped Flange, Z, Inches
$\frac{1}{2}$	$\frac{9}{16}$	$\frac{7}{8}$	$\frac{7}{8}$
$\frac{3}{4}$	$\frac{5}{8}$	1	1
1	$\frac{11}{16}$	$1\frac{1}{16}$	$1\frac{1}{16}$
$1\frac{1}{4}$	$\frac{13}{16}$	$1\frac{1}{8}$	$1\frac{1}{8}$
$1\frac{1}{2}$	$\frac{7}{8}$	$1\frac{1}{4}$	$1\frac{1}{4}$
2	1	$1\frac{7}{16}$	$1\frac{7}{16}$
$2\frac{1}{2}$	$1\frac{1}{8}$	$1\frac{5}{8}$	$1\frac{5}{8}$
3	$1\frac{1}{4}$	$1\frac{13}{16}$	$1\frac{13}{16}$
$3\frac{1}{2}$	$1\frac{3}{8}$	$1\frac{15}{16}$	$1\frac{15}{16}$
4	$1\frac{3}{8}$	2	2
5	$1\frac{1}{2}$	$2\frac{1}{8}$	$2\frac{1}{8}$
6	$1\frac{5}{8}$	$2\frac{1}{4}$	$2\frac{1}{4}$
8	$1\frac{7}{8}$	$2\frac{11}{16}$	$2\frac{11}{16}$
10	$2\frac{1}{8}$	$2\frac{7}{8}$	4
12	$2\frac{1}{4}$	$3\frac{1}{8}$	$4\frac{1}{4}$
14 OD	$2\frac{3}{8}$	$3\frac{5}{16}$	$4\frac{5}{8}$
16 OD	$2\frac{1}{2}$	$3\frac{11}{16}$	5
18 OD	$2\frac{5}{8}$	$3\frac{7}{8}$	$5\frac{3}{8}$
20 OD	$2\frac{3}{4}$	4	$5\frac{3}{4}$
24 OD	3	$4\frac{1}{2}$	$6\frac{1}{4}$

Laying Lengths of 400-Lb. Steel
Flanged Elbows, Tees and Crosses
(¼ Inch Raised Face Type)

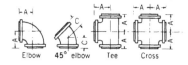

Elbow   45° elbow   Tee   Cross

Pipe Diameter, Inches	Dimension A, Inches	Dimension C, Inches
½	3¼	2
¾	3¾	2½
1	4¼	2½
1¼	4½	2¾
1½	4¾	3
2	5¾	4¼
2½	6½	4½
3	7	5
3½	7½	5½
4	8	5½
5	9	6
6	9¾	6¼
8	11¾	6¾
10	13¼	7¾
12	15	8¾
14 OD	16¼	9¼
16 OD	17¾	10¼
18 OD	19¼	10¾
20 OD	20¾	11¼
24 OD	24¼	12¾

Laying Lengths of 400-Lb. Steel Flanged Laterals and Reducers
(¼ Inch Raised Face Type)

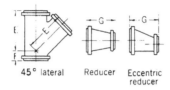

45° lateral   Reducer   Eccentric reducer

Pipe Diameter, Inches	Dimension E, Inches	Dimension F, Inches	Dimension G, Inches
½	5¾	1¾	5
¾	6¾	2	5
1	7¼	2¼	5
1¼	8	2½	5
1½	9	2¾	5
2	10¼	3½	6
2½	11½	3½	6¾
3	12¾	4	7¼
3½	14	4½	7¾
4	16	4½	8¼
5	16¾	5	9¼
6	18¾	5¼	10
8	22¼	5¾	12
10	25¾	6¼	13½
12	29¾	6½	15¼
14 OD	32¾	7	16½
16 OD	36¼	8	18½
18 OD	39¼	8½	19½
20 OD	42¾	9	21
24 OD	50¼	10½	24½

Laying Lengths of 400-Lb. Steel Flanged Elbows, Tees and Crosses

(Ring Joint Type)

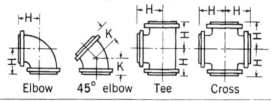

Elbow   45° elbow   Tee   Cross

Pipe Diameter, Inches	Dimension H, Inches	Dimension K, Inches
½	3³⁄₁₆	1¹⁵⁄₁₆
¾	3¾	2½
1	4¼	2½
1¼	4½	2¾
1½	4¾	3
2	5¹³⁄₁₆	4⁵⁄₁₆
2½	6⁹⁄₁₆	4⁹⁄₁₆
3	7¹⁄₁₆	5¹⁄₁₆
3½	7⁹⁄₁₆	5⁹⁄₁₆
4	8¹⁄₁₆	5⁹⁄₁₆
5	9¹⁄₁₆	6¹⁄₁₆
6	9¹³⁄₁₆	6⁵⁄₁₆
8	11¹³⁄₁₆	6¹³⁄₁₆
10	13⁵⁄₁₆	7¹³⁄₁₆
12	15¹⁄₁₆	8¹³⁄₁₆
14 OD	16⁵⁄₁₆	9⁵⁄₁₆
16 OD	17¹³⁄₁₆	10⁵⁄₁₆
18 OD	19⁵⁄₁₆	10¹³⁄₁₆
20 OD	20⅞	11⅜
24 OD	24⁷⁄₁₆	12¹⁵⁄₁₆

Laying Lengths of 400-Lb. Steel Flanged Laterals and Reducers

(Ring Joint Type)

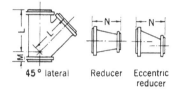

45° lateral   Reducer   Eccentric reducer

Pipe Diameter, Inches	Dimension L, Inches	Dimension M, Inches	Dimension N, Inches
$\frac{1}{2}$	$5\frac{11}{16}$	$1\frac{11}{16}$	$4\frac{7}{8}$
$\frac{3}{4}$	$6\frac{3}{4}$	2	$4\frac{15}{16}$
1	$7\frac{1}{4}$	$2\frac{1}{4}$	$4\frac{15}{16}$
$1\frac{1}{4}$	8	$2\frac{1}{2}$	$4\frac{15}{16}$
$1\frac{1}{2}$	9	$2\frac{3}{4}$	$4\frac{15}{16}$
2	$10\frac{5}{16}$	$3\frac{9}{16}$	$6\frac{1}{8}$
$2\frac{1}{2}$	$11\frac{9}{16}$	$3\frac{9}{16}$	$6\frac{7}{8}$
3	$12\frac{13}{16}$	$4\frac{1}{16}$	$7\frac{3}{8}$
$3\frac{1}{2}$	$14\frac{1}{16}$	$4\frac{9}{16}$	$7\frac{7}{8}$
4	$16\frac{1}{16}$	$4\frac{9}{16}$	$8\frac{3}{8}$
5	$16\frac{13}{16}$	$5\frac{1}{16}$	$9\frac{3}{8}$
6	$18\frac{13}{16}$	$5\frac{5}{16}$	$10\frac{1}{8}$
8	$22\frac{5}{16}$	$5\frac{13}{16}$	$12\frac{1}{8}$
10	$25\frac{13}{16}$	$6\frac{5}{16}$	$13\frac{5}{8}$
12	$29\frac{13}{16}$	$6\frac{9}{16}$	$15\frac{3}{8}$
14 OD	$32\frac{13}{16}$	$7\frac{1}{16}$	$16\frac{5}{8}$
16 OD	$36\frac{5}{16}$	$8\frac{1}{16}$	$18\frac{5}{8}$
18 OD	$39\frac{5}{16}$	$8\frac{9}{16}$	$19\frac{5}{8}$
20 OD	$42\frac{7}{8}$	$9\frac{1}{8}$	$21\frac{1}{4}$
24 OD	$50\frac{7}{16}$	$10\frac{11}{16}$	$24\frac{7}{8}$

Laying Lengths of 600-Lb. Steel Flanges

Blind

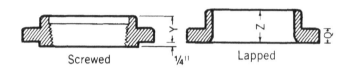

Screwed   Lapped

Pipe Diameter, Inches	Flange Thickness, Q, Inches	Screwed Flange, Y, Inches	Lapped Flange, Z, Inches
$\frac{1}{2}$	$\frac{9}{16}$	$\frac{7}{8}$	$\frac{7}{8}$
$\frac{3}{4}$	$\frac{5}{8}$	1	1
1	$\frac{11}{16}$	$1\frac{1}{16}$	$1\frac{1}{16}$
$1\frac{1}{4}$	$\frac{13}{16}$	$1\frac{1}{8}$	$1\frac{1}{8}$
$1\frac{1}{2}$	$\frac{7}{8}$	$1\frac{1}{4}$	$1\frac{1}{4}$
2	1	$1\frac{7}{16}$	$1\frac{7}{16}$
$2\frac{1}{2}$	$1\frac{1}{8}$	$1\frac{5}{8}$	$1\frac{5}{8}$
3	$1\frac{1}{4}$	$1\frac{13}{16}$	$1\frac{13}{16}$
$3\frac{1}{2}$	$1\frac{3}{8}$	$1\frac{15}{16}$	$1\frac{15}{16}$
4	$1\frac{1}{2}$	$2\frac{1}{8}$	$2\frac{1}{8}$
5	$1\frac{3}{4}$	$2\frac{3}{8}$	$2\frac{3}{8}$
6	$1\frac{7}{8}$	$2\frac{5}{8}$	$2\frac{5}{8}$
8	$2\frac{3}{16}$	3	3
10	$2\frac{1}{2}$	$3\frac{3}{8}$	$4\frac{3}{8}$
12	$2\frac{5}{8}$	$3\frac{5}{8}$	$4\frac{5}{8}$
14 OD	$2\frac{3}{4}$	$3\frac{11}{16}$	5
16 OD	3	$4\frac{3}{16}$	$5\frac{1}{2}$
18 OD	$3\frac{1}{4}$	$4\frac{5}{8}$	6
20 OD	$3\frac{1}{2}$	5	$6\frac{1}{2}$
24 OD	4	$5\frac{1}{2}$	$7\frac{1}{4}$

Laying Lengths of 600-Lb. Steel
Flanged Elbows, Tees and Crosses
(¼ Inch Raised Face Type)

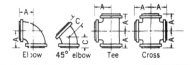

Elbow   45° elbow   Tee   Cross

Pipe Diameter, Inches	Dimension A, Inches	Dimension C, Inches
½	3¼	2
¾	3¾	2½
1	4¼	2½
1¼	4½	2¾
1½	4¾	3
2	5¾	4¼
2½	6½	4½
3	7	5
3½	7½	5½
4	8½	6
5	10	7
6	11	7½
8	13	8½
10	15½	9½
12	16½	10
14 OD	17½	10¾
16 OD	19½	11¾
18 OD	21½	12¼
20 OD	23½	13
24 OD	27½	14¾

Laying Lengths of 600-Lb. Steel
Flanged Elbows, Tees and Crosses
(Ring Joint Type)

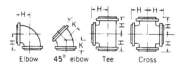

Elbow   45° elbow   Tee   Cross

Pipe Diameter, Inches	Dimension H, Inches	Dimension K, Inches
½	3³⁄₁₆	1¹⁵⁄₁₆
¾	3²³⁄₃₂	2¹⁵⁄₃₂
1	4⁷⁄₃₂	2¹⁵⁄₃₂
1¼	4¹⁵⁄₃₂	2²³⁄₃₂
1½	4²³⁄₃₂	2³¹⁄₃₂
2	5¹³⁄₁₆	4⁵⁄₁₆
2½	6⁹⁄₁₆	4⁹⁄₁₆
3	7¹⁄₁₆	5¹⁄₁₆
3½	7⁹⁄₁₆	5⁹⁄₁₆
4	8⁹⁄₁₆	6¹⁄₁₆
5	10¹⁄₁₆	7¹⁄₁₆
6	11¹⁄₁₆	7⁹⁄₁₆
8	13¹⁄₁₆	8⁹⁄₁₆
10	15⁹⁄₁₆	9⁹⁄₁₆
12	16⁹⁄₁₆	10¹⁄₁₆
14 OD	17⁹⁄₁₆	10¹³⁄₁₆
16 OD	19⁹⁄₁₆	11¹³⁄₁₆
18 OD	21⁹⁄₁₆	12⁵⁄₁₆
20 OD	23⅝	13⅛
24 OD	27¹¹⁄₁₆	14¹⁵⁄₁₆

Laying Lengths of 900-Lb. Steel Flanges

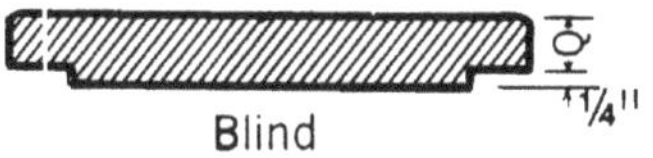

Blind

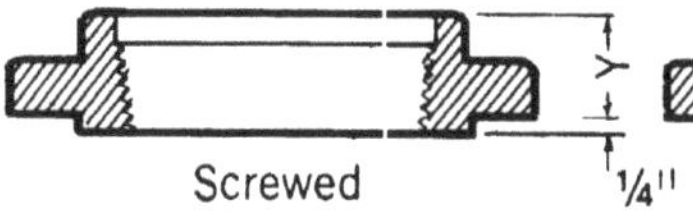

Screwed

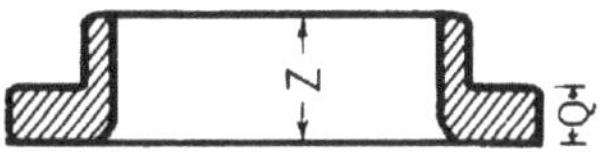

Lapped

Pipe Diameter, Inches	Flange Thickness, Q, Inches	Screwed Flange, Y, Inches	Lapped Flange, Z, Inches
1/2	7/8	1 1/4	1 1/4
3/4	1	1 3/8	1 3/8
1	1 1/8	1 5/8	1 5/8
1 1/4	1 1/8	1 5/8	1 5/8
1 1/2	1 1/4	1 3/4	1 3/4
2	1 1/2	2 1/4	2 1/4
2 1/2	1 5/8	2 1/2	2 1/2
3	1 1/2	2 1/8	2 1/8
4	1 3/4	2 3/4	2 3/4
5	2	3 1/8	3 1/8
6	2 3/16	3 3/8	3 3/8
8	2 1/2	4	4 1/2
10	2 3/4	4 1/4	5
12	3 1/8	4 5/8	5 5/8
14 OD	3 3/8	5 1/8	6 1/8
16 OD	3 1/2	5 1/4	6 1/2
18 OD	4	6	7 1/2
20 OD	4 1/4	6 1/4	8 1/4
24 OD	5 1/2	8	10 1/2

Laying Lengths of 900-Lb. Steel Flanged Elbows, Tees and Crosses
(¼ Inch Raised Face Type)

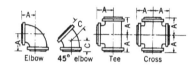

Elbow   45° elbow   Tee   Cross

Pipe Diameter, Inches	Dimension A, Inches	Dimension C, Inches
½	4¼	3
¾	4½	3¼
1	5	3½
1¼	5½	4
1½	6	4¼
2	7¼	4¾
2½	8¼	5¼
3	7½	5½
4	9	6½
5	11	7½
6	12	8
8	14½	9
10	16½	10
12	19	11
14 OD	20¼	11½
16 OD	22¼	12½
18 OD	24	13¼
20 OD	26	14½
24 OD	30½	18

Laying Lengths of 900-Lb. Steel Flanged Elbows, Tees and Crosses
(Ring Joint Type)

Elbow   45° elbow   Tee   Cross

Pipe Diameter, Inches	Dimension H, Inches	Dimension K, Inches
½	—	—
¾	—	—
1	$4\frac{31}{32}$	$3\frac{15}{32}$
1¼	$5\frac{15}{32}$	$3\frac{31}{32}$
1½	$5\frac{31}{32}$	$4\frac{7}{32}$
2	$7\frac{5}{16}$	$4\frac{13}{16}$
2½	$8\frac{5}{16}$	$5\frac{5}{16}$
3	$7\frac{9}{16}$	$5\frac{9}{16}$
4	$9\frac{1}{16}$	$6\frac{9}{16}$
5	$11\frac{1}{16}$	$7\frac{9}{16}$
6	$12\frac{1}{16}$	$8\frac{1}{16}$
8	$14\frac{9}{16}$	$9\frac{1}{16}$
10	$16\frac{9}{16}$	$10\frac{1}{16}$
12	$19\frac{1}{16}$	$11\frac{1}{16}$
14 OD	$20\frac{7}{16}$	$11\frac{11}{16}$
16 OD	$22\frac{7}{16}$	$12\frac{11}{16}$
18 OD	$24\frac{1}{4}$	$13\frac{1}{2}$
20 OD	$26\frac{1}{4}$	$14\frac{3}{4}$
24 OD	$30\frac{7}{8}$	$18\frac{3}{8}$

Laying Lengths of
1500-Lb. Steel Flanges

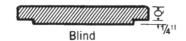

Blind

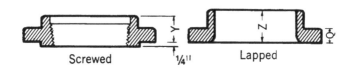

Screwed   ¼"   Lapped

Pipe Diameter, Inches	Flange Thickness, Q, Inches	Screwed Flange, Y, Inches	Lapped Flange, Z, Inches
½	⅞	1¼	1¼
¾	1	1⅜	1⅜
1	1⅛	1⅝	1⅝
1¼	1⅛	1⅝	1⅝
1½	1¼	1¾	1¾
2	1½	2¼	2¼
2½	1⅝	2½	2½
3	1⅞	2⅞	2⅞
4	2⅛	3⁹⁄₁₆	3⁹⁄₁₆
5	2⅞	4⅛	4⅛
6	3¼	4¹¹⁄₁₆	4¹¹⁄₁₆
8	3⅝	5⅝	5⅝
10	4¼	6¼	7
12	4⅞	7⅛	8⅝
14 OD	5¼	—	9½
16 OD	5¾	—	10¼
18 OD	6⅜	—	10⅞
20 OD	7	—	11½
24 OD	8	—	13

Laying Lengths of 1500-Lb. Steel Flanged Elbows, Tees and Crosses
(¼ Inch Raised Face Type)

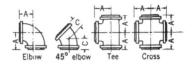

Elbow   45° elbow   Tee   Cross

Pipe Diameter, Inches	Dimension A, Inches	Dimension C, Inches
½	4¼	3
¾	4½	3¼
1	5	3½
1¼	5½	4
1½	6	4¼
2	7¼	4¾
2½	8¼	5¼
3	9¼	5¾
4	10¾	7¼
5	13¼	8¾
6	13⅞	9⅜
8	16⅜	10⅞
10	19½	12
12	22¼	13¼
14 OD	24¾	14¼
16 OD	27¼	16¼
18 OD	30¼	17¾
20 OD	32¾	18¾
24 OD	38¼	20¾

Laying Lengths of 1500-Lb. Steel Flanged Elbows, Tees and Crosses

(Ring Joint Type)

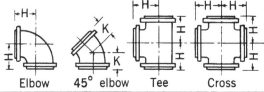

Elbow   45° elbow   Tee   Cross

Pipe Diameter, Inches	Dimension H, Inches	Dimension K, Inches
$\frac{1}{2}$	—	—
$\frac{3}{4}$	—	—
1	$4\frac{31}{32}$	$3\frac{15}{32}$
$1\frac{1}{4}$	$5\frac{15}{32}$	$3\frac{31}{32}$
$1\frac{1}{2}$	$5\frac{31}{32}$	$4\frac{7}{32}$
2	$7\frac{5}{16}$	$4\frac{13}{16}$
$2\frac{1}{2}$	$8\frac{5}{16}$	$5\frac{5}{16}$
3	$9\frac{5}{16}$	$5\frac{13}{16}$
4	$10\frac{13}{16}$	$7\frac{5}{16}$
5	$13\frac{5}{16}$	$3\frac{13}{16}$
6	14	$9\frac{1}{2}$
8	$16\frac{9}{16}$	$11\frac{1}{16}$
10	$19\frac{11}{16}$	$12\frac{3}{16}$
12	$22\frac{9}{16}$	$13\frac{9}{16}$
14 OD	$25\frac{1}{8}$	$14\frac{5}{8}$
16 OD	$27\frac{11}{16}$	$16\frac{11}{16}$
18 OD	$30\frac{11}{16}$	$18\frac{3}{16}$
20 OD	$33\frac{3}{16}$	$19\frac{3}{16}$
24 OD	$38\frac{13}{16}$	$21\frac{5}{16}$

Laying Lengths of
2500-Lb. Steel Flanges

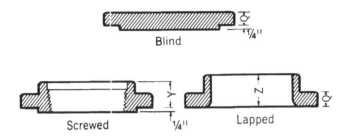

Blind

Screwed ¼" Lapped

Pipe Diameter, Inches	Flange Thickness, Q, Inches	Screwed Flange, Y, Inches	Lapped Flange, Z, Inches
½	1³⁄₁₆	1⁹⁄₁₆	1⁹⁄₁₆
¾	1¼	1¹¹⁄₁₆	1¹¹⁄₁₆
1	1⅜	1⅞	1⅞
1¼	1½	2¹⁄₁₆	2¹⁄₁₆
1½	1¾	2⅜	2⅜
2	2	2¾	2¾
2½	2¼	3⅛	3⅛
3	2⅝	3⅝	3⅝
4	3	4¼	4¼
5	3⅝	5⅛	5⅛
6	4¼	6	6
8	5	7	7
10	6½	9	9
12	7¼	10	10

Laying Lengths of 2500-Lb. Steel
Flanged Elbows, Tees and Crosses
(¼ Inch Raised Face Type)

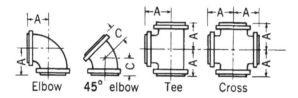

Elbow   45° elbow   Tee   Cross

Pipe Diameter, Inches	Dimension A, Inches	Dimension C, Inches
½	5³⁄₁₆	—
¾	5⅜	—
1	6¹⁄₁₆	4
1¼	6⅞	4¼
1½	7⁹⁄₁₆	4¾
2	8⅞	5¾
2½	10	6¼
3	11⅜	7¼
4	13¼	8½
5	15⅝	10
6	18	11½
8	20⅛	12¾
10	25	16
12	28	17¾

Laying Lengths of
2500-Lb. Steel
Flanged Elbows, Tees and Crosses
(Ring Joint Type)

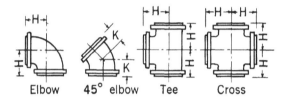

Elbow   45° elbow   Tee   Cross

Pipe Diameter, Inches	Dimension H, Inches	Dimension K, Inches
½	$5\frac{5}{32}$	—
¾	$5\frac{11}{32}$	—
1	$6\frac{1}{32}$	$3\frac{31}{32}$
1¼	$6\frac{15}{16}$	$4\frac{5}{16}$
1½	$7\frac{5}{8}$	$4\frac{13}{16}$
2	$8\frac{15}{16}$	$5\frac{13}{16}$
2½	$10\frac{1}{8}$	$6\frac{3}{8}$
3	$11\frac{1}{2}$	$7\frac{3}{8}$
4	$13\frac{7}{16}$	$8\frac{11}{16}$
5	$15\frac{7}{8}$	$10\frac{1}{4}$
6	$18\frac{1}{4}$	$11\frac{3}{4}$
8	$20\frac{7}{16}$	$13\frac{1}{16}$
10	$25\frac{7}{16}$	$16\frac{7}{16}$
12	$28\frac{7}{16}$	$18\frac{3}{16}$

Laying Lengths of
Cast Iron Flanged Wedge Gate Valves

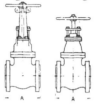

Nominal Size, Inches	Pressure, Lb. per Sq. In.		
	125	**175**	**250**
	Dimension A, Inches		
1	—	—	—
1¼	—	—	—
1½	—	—	—
2	7	7¼	8½
2½	7½	8	9½
3	8	9¼	11⅛
3½	8½	10	11⅞
4	9	10½	12
5	10	11½	15
6	10½	13	15⅞
8	11½	14¼	16½
10	13	16¾	18
12	14	17½	19¾
14 OD	15	—	22½
16 OD	16	—	24
18 OD	17	—	26
20 OD	18	—	28
24 OD	20	—	31

Laying Lengths of
Cast Iron Double Disc Flanged Gate Valves

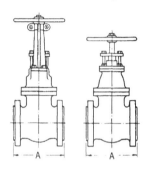

Nominal Size, Inches	Pressure, Lb. per Sq. In.		
	125	175	250
	Dimension A, Inches		
2	7	7¼	8½
2½	7½	8	9½
3	8	9¼	11⅛
3½	8½	—	—
4	9	10½	12
5	10	—	—
6	10½	13	15⅞
8	11½	14¼	16½
10	13	16¾	18
12	14	17½	19¾
14 OD	—	—	22½
16 OD	—	—	24
18 OD	—	—	26
20 OD	—	—	28
24 OD	—	—	31

Laying Lengths of
Steel Flanged Wedge Gate Valves

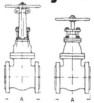

Nominal Size, Inches	Pressure, Lb. per Sq. In.			
	150	300	400	600
	Dimension A, Inches			
1	—	—	$8\frac{1}{2}$	$8\frac{1}{2}$
$1\frac{1}{4}$	—	—	9	9
$1\frac{1}{2}$	—	$7\frac{1}{2}$	$9\frac{1}{2}$	$9\frac{1}{2}$
2	7	$8\frac{1}{2}$	$11\frac{1}{2}$	$11\frac{1}{2}$
$2\frac{1}{2}$	$7\frac{1}{2}$	$9\frac{1}{2}$	13	13
3	8	$11\frac{1}{8}$	14	14
$3\frac{1}{2}$	$8\frac{1}{2}$	$11\frac{7}{8}$	—	—
4	9	12	16	17
5	10	15	18	20
6	$10\frac{1}{2}$	$15\frac{7}{8}$	$19\frac{1}{2}$	22
8	$11\frac{1}{2}$	$16\frac{1}{2}$	$23\frac{1}{2}$	26
10	13	18	$26\frac{1}{2}$	31
12	14	$19\frac{3}{4}$	30	33
14 OD	15	30	$32\frac{1}{2}$	35
16 OD	16	33	$35\frac{1}{2}$	39
18 OD	17	36	$38\frac{1}{2}$	43
20 OD	18	39	$41\frac{1}{2}$	47
24 OD	20	45	$48\frac{1}{2}$	55

Laying Lengths of
Steel Flanged Solid Wedge and Double
Disc Gate Valves — Raised Face Type

Nominal Size, Inches	Pressure, Lb. per Sq. In.		
	900	1500	2500
	Dimension A, Inches		
1	10	10	$12\frac{1}{8}$
$1\frac{1}{4}$	11	11	$13\frac{3}{4}$
$1\frac{1}{2}$	12	12	$15\frac{1}{8}$
2	$14\frac{1}{2}$	$14\frac{1}{2}$	$17\frac{3}{4}$
$2\frac{1}{2}$	$16\frac{1}{2}$	$16\frac{1}{2}$	20
3	15	$18\frac{1}{2}$	$22\frac{3}{4}$
4	18	$21\frac{1}{2}$	$26\frac{1}{2}$
5	22	$26\frac{1}{2}$	$31\frac{1}{4}$
6	24	$27\frac{3}{4}$	36
8	29	$32\frac{3}{4}$	$40\frac{1}{4}$
10	33	39	50
12	38	$44\frac{1}{2}$	56
14	$40\frac{1}{2}$	$49\frac{1}{2}$	—
16	$44\frac{1}{2}$	$54\frac{1}{2}$	—
18	48	$60\frac{1}{2}$	—
20	52	$65\frac{1}{2}$	—
24	61	$76\frac{1}{2}$	—

Laying Lengths of
Steel Flanged Solid Wedge and Double
Disc Gate Valves — Ring Joint Type

Nominal Size, Inches	Pressure, Lb. per Sq. In.			
	150	300	400	600
	Dimension A, Inches			
1	5½	—	8½	8½
1¼	6	—	9	9
1½	7	8	9½	9½
2	7½	9⅛	11⅝	11⅝
2½	8	10⅛	13⅛	13⅛
3	8½	11¾	14⅛	14⅛
4	9½	12⅝	16⅛	17⅛
5	10½	15⅝	18⅛	20⅛
6	11	16½	19⅝	22⅛
8	12	17⅛	23⅝	26⅛
10	13½	18⅝	26⅝	31⅛
12	14½	20⅜	30⅛	33⅛
14	15½	30⅝	32⅝	35⅛
16	16½	33⅝	35⅝	39⅛
18	17½	36⅝	38⅝	43⅛
20	18½	39¾	41¾	47¼
24	20½	45⅞	48⅞	55⅜

Laying Lengths of
Steel Flanged Solid Wedge and Double
Disc Gate Valves — Ring Joint Type

Nominal Size, Inches	Pressure, Lb. per Sq. In.		
	900	1500	2500
	Dimension A, Inches		
1	10	10	$12\frac{1}{8}$
$1\frac{1}{4}$	11	11	$13\frac{7}{8}$
$1\frac{1}{2}$	12	12	$15\frac{1}{4}$
2	$14\frac{5}{8}$	$14\frac{5}{8}$	$17\frac{7}{8}$
$2\frac{1}{2}$	$16\frac{5}{8}$	$16\frac{5}{8}$	$20\frac{1}{4}$
3	$15\frac{1}{8}$	$18\frac{5}{8}$	23
4	$18\frac{1}{8}$	$21\frac{5}{8}$	$26\frac{7}{8}$
5	$22\frac{1}{8}$	$26\frac{5}{8}$	$31\frac{3}{4}$
6	$24\frac{1}{8}$	28	$36\frac{1}{2}$
8	$29\frac{1}{8}$	$33\frac{1}{8}$	$40\frac{7}{8}$
10	$33\frac{1}{8}$	$39\frac{3}{8}$	$50\frac{7}{8}$
12	$38\frac{1}{8}$	$45\frac{1}{8}$	$56\frac{7}{8}$
14	$40\frac{7}{8}$	$50\frac{1}{4}$	—
16	$44\frac{7}{8}$	$55\frac{3}{8}$	—
18	$48\frac{1}{2}$	$61\frac{3}{8}$	—
20	$52\frac{1}{2}$	$66\frac{3}{8}$	—
24	$61\frac{3}{4}$	$77\frac{5}{8}$	—

Laying Lengths of
Cast Iron Flanged Globe and Angle Valves

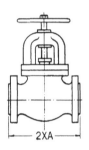
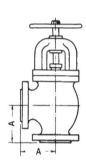

Nominal Size, Inches	Pressure, Lb. per Sq. In.	
	125	250
	Dimension 2 x A, Inches	
½	—	—
¾	—	—
1	—	—
1¼	—	—
1½	—	—
2	8	10½
2½	8½	11½
3	9½	12½
3½	10½	13¼
4	11½	14
5	13	15¾
6	14	17½
8	19½	21

Laying Lengths of
Steel Flanged Globe and Angle Valves
Raised Face Type

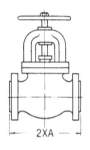 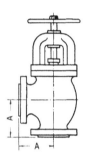

Nominal Size, Inches	Pressure, Lb. per Sq. In.			
	150	300	400	600
	Dimension 2 x A, Inches			
½	—	—	—	—
¾	—	—	7½	7½
1	—	—	8½	8½
1¼	—	—	9	9
1½	—	—	9½	9½
2	8	10½	11½	11½
2½	8½	11½	13	13
3	9½	12½	14	14
3½	10½	13¼	—	—
4	11½	14	16	17
5	14	15¾	18	20
6	16	17½	19½	22
8	19½	22	23½	26

Laying Lengths of
Steel Flanged Globe and Angle Valves
Raised Face Type

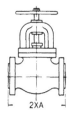 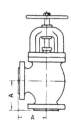

Nominal Size, Inches	Pressure, Lb. per Sq. In.		
	900	1500	2500
	Dimension 2 x A, Inches		
½	—	—	10⅜
¾	9	9	10¾
1	10	10	12⅛
1¼	11	11	13¾
1½	12	12	15⅛
2	14½	14½	17¾
2½	16½	16½	20
3	15	18½	22¾
4	18	21½	26½
5	22	26½	31¼
6	24	27¾	36
8	29	32¾	40¼
10	33	39	50
12	38	44½	56
14	40½	49½	—

Laying Lengths of
Steel Flanged Globe and Angle Valves
Ring Joint Type

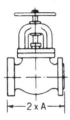 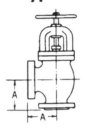

Nominal Size, Inches	Pressure, Lb. per Sq. In.			
	150	300	400	600
	Dimension 2 x A, Inches			
½	—	6$\frac{7}{16}$	6$\frac{7}{16}$	6$\frac{7}{16}$
¾	—	7½	7½	7½
1	—	8½	8½	8½
1¼	—	9	9	9
1½	7	9½	9½	9½
2	8½	11$\frac{1}{8}$	11$\frac{5}{8}$	11$\frac{5}{8}$
2½	9	12$\frac{1}{8}$	13$\frac{1}{8}$	13$\frac{1}{8}$
3	—	13$\frac{1}{8}$	14$\frac{1}{8}$	14$\frac{1}{8}$
4	12	14$\frac{5}{8}$	16$\frac{1}{8}$	17$\frac{1}{8}$
5	14½	16$\frac{3}{8}$	18$\frac{1}{8}$	20$\frac{1}{8}$
6	16½	18$\frac{1}{8}$	19$\frac{5}{8}$	22$\frac{1}{8}$
8	20	22$\frac{5}{8}$	23$\frac{5}{8}$	26$\frac{1}{8}$
10	25	25$\frac{1}{8}$	26$\frac{5}{8}$	31$\frac{1}{8}$
12	28	28$\frac{5}{8}$	30$\frac{1}{8}$	33$\frac{1}{8}$
14	31½	—	—	—
16	36½	—	—	—

Laying Lengths of
Steel Flanged Globe and Angle Valves
Ring Joint Type

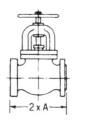 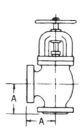

Nominal Size, Inches	Pressure, Lb. per Sq. In.		
	900	**1500**	**2500**
	Dimension 2 x A, Inches		
$\frac{1}{2}$	—	—	$10\frac{3}{8}$
$\frac{3}{4}$	9	9	$10\frac{3}{4}$
1	10	10	$12\frac{1}{8}$
$1\frac{1}{4}$	11	11	$13\frac{7}{8}$
$1\frac{1}{2}$	12	12	$15\frac{1}{4}$
2	$14\frac{5}{8}$	$14\frac{5}{8}$	$17\frac{7}{8}$
$2\frac{1}{2}$	$16\frac{5}{8}$	$16\frac{5}{8}$	$20\frac{1}{4}$
3	$15\frac{1}{8}$	$18\frac{5}{8}$	23
4	$18\frac{1}{8}$	$21\frac{5}{8}$	$26\frac{7}{8}$
5	$22\frac{1}{8}$	$26\frac{5}{8}$	$31\frac{3}{4}$
6	$24\frac{1}{8}$	28	$36\frac{1}{2}$
8	$29\frac{1}{8}$	$33\frac{1}{8}$	$40\frac{7}{8}$
10	$33\frac{1}{8}$	$39\frac{3}{8}$	$50\frac{7}{8}$
12	$38\frac{1}{8}$	$45\frac{1}{8}$	$56\frac{7}{8}$
14	$40\frac{7}{8}$	$50\frac{1}{4}$	—

Laying Lengths of
Cast Iron Flanged Swing Check Valves

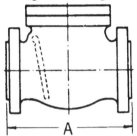

Where dimensions are not given the sizes either are not made or there is insufficient demand to warrant the expense of unification. Female or groove joint faces have bottom of groove in same place as "flange edge", and center to contact surface dimensions for these faces are reduced by the amount of the raised face. Dimensions given in the table are not intended to cover the type of check valve having the seat angle at approximately 45 degrees to the run of the valve, or other patterns where large clearances are required.

Nominal Size, Inches	Pressure, Lb. per Sq. In.	
	125	250
	Dimension A, Inches	
2	8	10½
2½	8½	11½
3	9½	12½
3½	10½	13¼
4	11½	14
5	13	15¾
6	14	17½
8	—	21
10	—	24½
12	—	28

Laying Lengths of
Steel Flanged Swing Check Valves

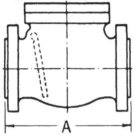

Where dimensions are not given the sizes either are not made or there is insufficient demand to warrant the expense of unification. Female and groove joint faces have bottom of groove in same place as "flange edge", and center to contact surface dimensions for these faces are reduced by the amount of the raised face. Dimensions given in the table are not intended to cover the type of check valve having the seat angle at approximately 45 degrees to the run of the valve or other patterns where large clearances are required.

Nominal Size, Inches	Pressure, Lb. per Sq. In.			
	150	300	400	600
	Dimension A, Inches			
2	8	10½	11½	11½
2½	8½	11½	13	13
3	9½	12½	14	14
3½	10½	13¼	—	—
4	11½	14	16	17
5	13	15¾	—	—
6	14	17½	19½	22
8	—	21	23½	26
10	—	24½	26½	31
12	—	28	30	33

Laying Lengths of
Steel Flanged Swing Check Valves
Raised Face Type

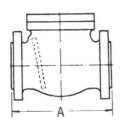

Nominal Size, Inches	Pressure, Lb. per Sq. In.		
	900	1500	2500
	Dimension A, Inches		
½	—	—	10⅜
¾	9	9	10¾
1	10	10	12⅛
1¼	11	11	13¾
1½	12	12	15⅛
2	14½	14½	17¾
2½	16½	16½	20
3	15	18½	22¾
4	18	21½	26½
5	22	26½	31¼
6	24	27¾	36
8	29	32¾	40¼
10	33	39	50
12	38	44½	56
14	40½	49½	—

Laying Lengths of
Steel Flanged Swing Check Valves
Ring Joint Type

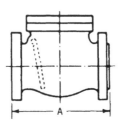

Nominal Size, Inches	Pressure, Lb. per Sq. In.			
	150	300	400	600
	Dimension A, Inches			
½	4¹¹⁄₁₆	—	6⁷⁄₁₆	6⁷⁄₁₆
¾	5⅛	—	7½	7½
1	5½	9	8½	8½
1¼	6	9½	9	9
1½	7	10	9½	9½
2	8½	11⅛	11⅝	11⅝
2½	9	12⅛	13⅛	13⅛
3	10	13⅛	14⅛	14⅛
4	12	14⅝	16⅛	17⅛
5	13½	16⅜	18⅛	20⅛
6	14½	18⅛	19⅝	22⅛
8	20	21⅝	23⅝	26⅛
10	25	25⅛	26⅝	31⅛
12	28	28⅝	30⅛	33⅛
14	31½	—	—	—

Laying Lengths of Steel Flanged Swing Check Valves Ring Joint Type

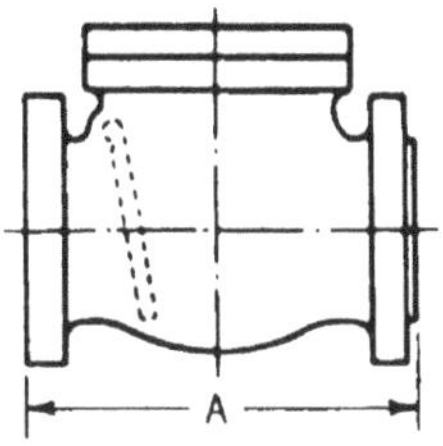

Nominal Size, Inches	Pressure, Lb. per Sq. In.		
	900	1500	2500
	Dimension A, Inches		
½	—	—	10⅜
¾	9	9	10¾
1	10	10	12⅛
1¼	11	11	13⅞
1½	12	12	15¼
2	14⅝	14⅝	17⅞
2½	16⅝	16⅝	20¼
3	15⅛	18⅝	23
4	18⅛	21⅝	26⅞
5	22⅛	26⅝	31¾
6	24⅛	28	36½
8	29⅛	33⅛	40⅞
10	33⅛	39⅜	50⅞
12	38⅛	45⅛	56⅞
14	40⅞	50¼	—

Laying Length of
Connecting Ring Joint Flanges

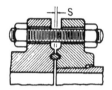

Length S is approximate distance between connecting flanges having octagonal or oval ring gaskets when rings are compressed.

Nominal Size, Inches	Pressure, Lb. per Sq. In.			
	150	300	400	600
	Dimension S, Inches			
$\frac{1}{2}$	$\frac{1}{8}$	$\frac{1}{8}$	$\frac{1}{8}$	$\frac{1}{8}$
$\frac{3}{4}$	$\frac{5}{32}$	$\frac{5}{32}$	$\frac{5}{32}$	$\frac{5}{32}$
1	$\frac{5}{32}$	$\frac{5}{32}$	$\frac{5}{32}$	$\frac{5}{32}$
$1\frac{1}{4}$	$\frac{5}{32}$	$\frac{5}{32}$	$\frac{5}{32}$	$\frac{5}{32}$
$1\frac{1}{2}$	$\frac{5}{32}$	$\frac{5}{32}$	$\frac{5}{32}$	$\frac{5}{32}$
2	$\frac{5}{32}$	$\frac{7}{32}$	$\frac{3}{16}$	$\frac{3}{16}$
$2\frac{1}{2}$	$\frac{5}{32}$	$\frac{7}{32}$	$\frac{3}{16}$	$\frac{3}{16}$
3	$\frac{5}{32}$	$\frac{7}{32}$	$\frac{3}{16}$	$\frac{3}{16}$
4	$\frac{5}{32}$	$\frac{7}{32}$	$\frac{7}{32}$	$\frac{3}{16}$
5	$\frac{5}{32}$	$\frac{7}{32}$	$\frac{7}{32}$	$\frac{3}{16}$
6	$\frac{5}{32}$	$\frac{7}{32}$	$\frac{7}{32}$	$\frac{3}{16}$
8	$\frac{5}{32}$	$\frac{7}{32}$	$\frac{7}{32}$	$\frac{3}{16}$
10	$\frac{5}{32}$	$\frac{7}{32}$	$\frac{7}{32}$	$\frac{3}{16}$
12	$\frac{5}{32}$	$\frac{7}{32}$	$\frac{7}{32}$	$\frac{3}{16}$
14	$\frac{1}{8}$	$\frac{7}{32}$	$\frac{7}{32}$	$\frac{3}{16}$
16	$\frac{1}{8}$	$\frac{7}{32}$	$\frac{7}{32}$	$\frac{3}{16}$
18	$\frac{1}{8}$	$\frac{7}{32}$	$\frac{7}{32}$	$\frac{3}{16}$
20	$\frac{1}{8}$	$\frac{7}{32}$	$\frac{7}{32}$	$\frac{3}{16}$
22	—	$\frac{1}{4}$	$\frac{1}{4}$	$\frac{7}{32}$
24	$\frac{1}{8}$	$\frac{1}{4}$	$\frac{1}{4}$	$\frac{7}{32}$

Laying Length of
Connecting Ring Joint Flanges

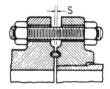

Length S is approximate distance between connecting flanges having octagonal or oval ring gaskets when rings are compressed.

Nominal Size, Inches	Pressure, Lb. per Sq. In.		
	900	1500	2500
	Dimension S, Inches		
$\frac{1}{2}$	—	—	$\frac{5}{32}$
$\frac{3}{4}$	$\frac{5}{32}$	$\frac{5}{32}$	$\frac{5}{32}$
1	$\frac{5}{32}$	$\frac{5}{32}$	$\frac{5}{32}$
$1\frac{1}{4}$	$\frac{5}{32}$	$\frac{5}{32}$	$\frac{1}{8}$
$1\frac{1}{2}$	$\frac{5}{32}$	$\frac{5}{32}$	$\frac{1}{8}$
2	$\frac{1}{8}$	$\frac{1}{8}$	$\frac{1}{8}$
$2\frac{1}{2}$	$\frac{1}{8}$	$\frac{1}{8}$	$\frac{1}{8}$
3	$\frac{5}{32}$	$\frac{1}{8}$	$\frac{1}{8}$
4	$\frac{5}{32}$	$\frac{1}{8}$	$\frac{5}{32}$
5	$\frac{5}{32}$	$\frac{1}{8}$	$\frac{5}{32}$
6	$\frac{5}{32}$	$\frac{1}{8}$	$\frac{5}{32}$
8	$\frac{5}{32}$	$\frac{5}{32}$	$\frac{3}{16}$
10	$\frac{5}{32}$	$\frac{5}{32}$	$\frac{1}{4}$
12	$\frac{5}{32}$	$\frac{3}{16}$	$\frac{5}{16}$
14	$\frac{5}{32}$	$\frac{7}{32}$	—
16	$\frac{5}{32}$	$\frac{5}{16}$	—
18	$\frac{3}{16}$	$\frac{5}{16}$	—
20	$\frac{3}{16}$	$\frac{3}{8}$	—
22	—	—	—
24	$\frac{7}{32}$	$\frac{7}{16}$	—

Dimensions of Ring Joint Gaskets

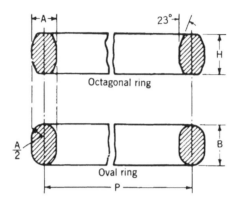

Ring Number	Pitch Diameter of Ring P, Inches	Width of Ring A, Inches	Height of Ring	
			Oval B, Inches	Octagonal H, Inches
R11	$1\frac{11}{32}$	$\frac{1}{4}$	$\frac{7}{16}$	$\frac{3}{8}$
R12	$1\frac{9}{16}$	$\frac{5}{16}$	$\frac{9}{16}$	$\frac{1}{2}$
R13	$1\frac{11}{16}$	$\frac{5}{16}$	$\frac{9}{16}$	$\frac{1}{2}$
R14	$1\frac{3}{4}$	$\frac{5}{16}$	$\frac{9}{16}$	$\frac{1}{2}$
R15	$1\frac{7}{8}$	$\frac{5}{16}$	$\frac{9}{16}$	$\frac{1}{2}$
R16	2	$\frac{5}{16}$	$\frac{9}{16}$	$\frac{1}{2}$
R17	$2\frac{1}{4}$	$\frac{5}{16}$	$\frac{9}{16}$	$\frac{1}{2}$
R18	$2\frac{3}{8}$	$\frac{5}{16}$	$\frac{9}{16}$	$\frac{1}{2}$
R19	$2\frac{9}{16}$	$\frac{5}{16}$	$\frac{9}{16}$	$\frac{1}{2}$
R20	$2\frac{11}{16}$	$\frac{5}{16}$	$\frac{9}{16}$	$\frac{1}{2}$
R21	$2\frac{27}{32}$	$\frac{7}{16}$	$\frac{11}{16}$	$\frac{5}{8}$
R22	$3\frac{1}{4}$	$\frac{5}{16}$	$\frac{9}{16}$	$\frac{1}{2}$
R23	$3\frac{1}{4}$	$\frac{7}{16}$	$\frac{11}{16}$	$\frac{5}{8}$
R24	$3\frac{3}{4}$	$\frac{7}{16}$	$\frac{11}{16}$	$\frac{5}{8}$
R25	4	$\frac{5}{16}$	$\frac{9}{16}$	$\frac{1}{2}$
R26	4	$\frac{7}{16}$	$\frac{11}{16}$	$\frac{5}{8}$

Dimensions of Ring Joint Gaskets
(Continued)

Ring Number	Pitch Diameter of Ring P, Inches	Width of Ring A, Inches	Height of Ring	
			Oval B, Inches	Octag-onal H, Inches
R27	$4\frac{1}{4}$	$\frac{7}{16}$	$\frac{11}{16}$	$\frac{5}{8}$
R28	$4\frac{3}{8}$	$\frac{1}{2}$	$\frac{3}{4}$	$\frac{11}{16}$
R29	$4\frac{1}{2}$	$\frac{5}{16}$	$\frac{9}{16}$	$\frac{1}{2}$
R30	$4\frac{5}{8}$	$\frac{7}{16}$	$\frac{11}{16}$	$\frac{5}{8}$
R31	$4\frac{7}{8}$	$\frac{7}{16}$	$\frac{11}{16}$	$\frac{5}{8}$
R32	5	$\frac{1}{2}$	$\frac{3}{4}$	$\frac{11}{16}$
R33	$5\frac{3}{16}$	$\frac{5}{16}$	$\frac{9}{16}$	$\frac{1}{2}$
R34	$5\frac{3}{16}$	$\frac{7}{16}$	$\frac{11}{16}$	$\frac{5}{8}$
R35	$5\frac{3}{8}$	$\frac{7}{16}$	$\frac{11}{16}$	$\frac{5}{8}$
R36	$5\frac{7}{8}$	$\frac{5}{16}$	$\frac{9}{16}$	$\frac{1}{2}$
R37	$5\frac{7}{8}$	$\frac{7}{16}$	$\frac{11}{16}$	$\frac{5}{8}$
R38	$6\frac{3}{16}$	$\frac{5}{8}$	$\frac{7}{8}$	$\frac{13}{16}$
R39	$6\frac{3}{8}$	$\frac{7}{16}$	$\frac{11}{16}$	$\frac{5}{8}$
R40	$6\frac{3}{4}$	$\frac{5}{16}$	$\frac{9}{16}$	$\frac{1}{2}$
R41	$7\frac{1}{8}$	$\frac{7}{16}$	$\frac{11}{16}$	$\frac{5}{8}$
R42	$7\frac{1}{2}$	$\frac{3}{4}$	1	$\frac{15}{16}$
R43	$7\frac{5}{8}$	$\frac{5}{16}$	$\frac{9}{16}$	$\frac{1}{2}$
R44	$7\frac{5}{8}$	$\frac{7}{16}$	$\frac{11}{16}$	$\frac{5}{8}$
R45	$8\frac{5}{16}$	$\frac{7}{16}$	$\frac{11}{16}$	$\frac{5}{8}$
R46	$8\frac{5}{16}$	$\frac{1}{2}$	$\frac{3}{4}$	$\frac{11}{16}$
R47	9	$\frac{3}{4}$	1	$\frac{15}{16}$
R48	$9\frac{3}{4}$	$\frac{5}{16}$	$\frac{9}{16}$	$\frac{1}{2}$
R49	$10\frac{5}{8}$	$\frac{7}{16}$	$\frac{11}{16}$	$\frac{5}{8}$
R50	$10\frac{5}{8}$	$\frac{5}{8}$	$\frac{7}{8}$	$\frac{13}{16}$
R51	11	$\frac{7}{8}$	$1\frac{1}{8}$	$1\frac{1}{16}$
R52	12	$\frac{5}{16}$	$\frac{9}{16}$	$\frac{1}{2}$

Dimensions of Ring Joint Gaskets

(Continued)

Ring Number	Pitch Diameter of Ring P, Inches	Width of Ring A, Inches	Height of Ring	
			Oval B, Inches	Octagonal H, Inches
R53	$12\frac{3}{4}$	$\frac{7}{16}$	$\frac{11}{16}$	$\frac{5}{8}$
R54	$12\frac{3}{4}$	$\frac{5}{8}$	$\frac{7}{8}$	$\frac{13}{16}$
R55	$13\frac{1}{2}$	$1\frac{1}{8}$	$1\frac{7}{16}$	$1\frac{3}{8}$
R56	15	$\frac{5}{16}$	$\frac{9}{16}$	$\frac{1}{2}$
R57	15	$\frac{7}{16}$	$\frac{11}{16}$	$\frac{5}{8}$
R58	15	$\frac{7}{8}$	$1\frac{1}{8}$	$1\frac{1}{16}$
R59	$15\frac{5}{8}$	$\frac{5}{16}$	$\frac{9}{16}$	$\frac{1}{2}$
R60	16	$1\frac{1}{4}$	$1\frac{9}{16}$	$1\frac{1}{2}$
R61	$16\frac{1}{2}$	$\frac{7}{16}$	$\frac{11}{16}$	$\frac{5}{8}$
R62	$16\frac{1}{2}$	$\frac{5}{8}$	$\frac{7}{8}$	$\frac{13}{16}$
R63	$16\frac{1}{2}$	1	$1\frac{5}{16}$	$1\frac{1}{4}$
R64	$17\frac{7}{8}$	$\frac{5}{16}$	$\frac{9}{16}$	$\frac{1}{2}$
R65	$18\frac{1}{2}$	$\frac{7}{16}$	$\frac{11}{16}$	$\frac{5}{8}$
R66	$18\frac{1}{2}$	$\frac{5}{8}$	$\frac{7}{8}$	$\frac{13}{16}$
R67	$18\frac{1}{2}$	$1\frac{1}{8}$	$1\frac{7}{16}$	$1\frac{3}{8}$
R68	$20\frac{3}{8}$	$\frac{5}{16}$	$\frac{9}{16}$	$\frac{1}{2}$
R69	21	$\frac{7}{16}$	$\frac{11}{16}$	$\frac{5}{8}$
R70	21	$\frac{3}{4}$	1	$\frac{15}{16}$
R71	21	$1\frac{1}{8}$	$1\frac{7}{16}$	$1\frac{3}{8}$
R72	22	$\frac{5}{16}$	$\frac{9}{16}$	$\frac{1}{2}$
R73	23	$\frac{1}{2}$	$\frac{3}{4}$	$\frac{11}{16}$
R74	23	$\frac{3}{4}$	1	$\frac{15}{16}$
R75	23	$1\frac{1}{4}$	$1\frac{9}{16}$	$1\frac{1}{2}$
R76	$26\frac{1}{2}$	$\frac{5}{16}$	$\frac{9}{16}$	$\frac{1}{2}$
R77	$27\frac{1}{4}$	$\frac{5}{8}$	$\frac{7}{8}$	$\frac{13}{16}$
R78	$27\frac{1}{4}$	1	$1\frac{5}{16}$	$1\frac{1}{4}$

Dimensions of Ring Joint Gaskets
(Continued)

Ring Number	Pitch Diameter of Ring P, Inches	Width of Ring A, Inches	Height of Ring — Oval B, Inches	Height of Ring — Octagonal H, Inches
R79	$27\frac{1}{4}$	$1\frac{3}{8}$	$1\frac{3}{4}$	$1\frac{5}{8}$
R80	$24\frac{1}{4}$	$\frac{5}{16}$	—	$\frac{1}{2}$
R81	25	$\frac{9}{16}$	—	$\frac{3}{4}$
R82	$2\frac{1}{4}$	$\frac{7}{16}$	—	$\frac{5}{8}$
R84	$2\frac{1}{2}$	$\frac{7}{16}$	—	$\frac{5}{8}$
R85	$3\frac{1}{8}$	$\frac{1}{2}$	—	$\frac{11}{16}$
R86	$3\frac{9}{16}$	$\frac{5}{8}$	—	$\frac{13}{16}$
R87	$3\frac{15}{16}$	$\frac{5}{8}$	—	$\frac{13}{16}$
R88	$4\frac{7}{8}$	$\frac{3}{4}$	—	$\frac{15}{16}$
R89	$4\frac{1}{2}$	$\frac{3}{4}$	—	$\frac{15}{16}$
R90	$6\frac{1}{8}$	$\frac{7}{8}$	—	$1\frac{1}{16}$
R91	$10\frac{1}{4}$	$1\frac{1}{4}$	—	$1\frac{1}{2}$
R92	9	$\frac{7}{16}$	$\frac{11}{16}$	$\frac{5}{8}$
R93	$29\frac{1}{2}$	$\frac{3}{4}$	—	$\frac{15}{16}$
R94	$31\frac{1}{2}$	$\frac{3}{4}$	—	$\frac{15}{16}$
R95	$33\frac{3}{4}$	$\frac{3}{4}$	—	$\frac{15}{16}$
R96	36	$\frac{7}{8}$	—	$1\frac{1}{16}$
R97	38	$\frac{7}{8}$	—	$1\frac{1}{16}$
R98	$40\frac{1}{4}$	$\frac{7}{8}$	—	$1\frac{1}{16}$
R99	$9\frac{1}{4}$	$\frac{7}{16}$	—	$\frac{5}{8}$
R100	$29\frac{1}{2}$	$1\frac{1}{8}$	—	$1\frac{3}{8}$
R101	$31\frac{1}{2}$	$1\frac{1}{4}$	—	$1\frac{1}{2}$
R102	$33\frac{3}{4}$	$1\frac{1}{4}$	—	$1\frac{1}{2}$
R103	36	$1\frac{1}{4}$	—	$1\frac{1}{2}$
R104	38	$1\frac{3}{8}$		$1\frac{5}{8}$
R105	$40\frac{1}{4}$	$1\frac{3}{8}$		$1\frac{5}{8}$

Dimensions of Flat Gaskets

Standard and Low Pressure Flanges*

Pipe Size, Inches	Ring Gaskets		Full Faced Gaskets	
	Inside Dia., Inches	Outside Dia., Inches	Inside Dia., Inches	Outside Dia., Inches
1	1	$2\frac{5}{8}$	1	$4\frac{1}{4}$
$1\frac{1}{4}$	$1\frac{1}{4}$	3	$1\frac{1}{4}$	$4\frac{5}{8}$
$1\frac{1}{2}$	$1\frac{1}{2}$	$3\frac{3}{8}$	$1\frac{1}{2}$	5
2	2	$4\frac{1}{8}$	2	6
$2\frac{1}{2}$	$2\frac{1}{2}$	$4\frac{7}{8}$	$2\frac{1}{2}$	7
3	3	$5\frac{3}{8}$	3	$7\frac{1}{2}$
4	4	$6\frac{7}{8}$	4	9
5	5	$7\frac{3}{4}$	5	10
6	6	$8\frac{3}{4}$	6	11
8	8	11	8	$13\frac{1}{2}$
10	10	$13\frac{3}{8}$	10	16
12	12	$16\frac{1}{8}$	12	19
14	14	$17\frac{3}{4}$	14	21
16	16	$20\frac{1}{4}$	16	$23\frac{1}{2}$
18	18	$21\frac{5}{8}$	18	25
20	20	$23\frac{7}{8}$	20	$27\frac{1}{2}$
24	24	$28\frac{1}{4}$	24	32

*Includes 25 lb. and 125 lb. Cast Iron; 150 lb. Steel flanges.

Dimensions of Flat Gaskets
(Continued)

Medium and Extra Heavy Flanges*

Pipe Size, Inches	Ring Gaskets		Full Faced Gaskets	
	Inside Dia., Inches	Outside Dia., Inches	Inside Dia., Inches	Outside Dia., Inches
1	1	$2\frac{7}{8}$	1	$4\frac{7}{8}$
$1\frac{1}{4}$	$1\frac{1}{4}$	$3\frac{1}{4}$	$1\frac{1}{4}$	$5\frac{1}{4}$
$1\frac{1}{2}$	$1\frac{1}{2}$	$3\frac{3}{4}$	$1\frac{1}{2}$	$6\frac{1}{8}$
2	2	$4\frac{3}{8}$	2	$6\frac{1}{2}$
$2\frac{1}{2}$	$2\frac{1}{2}$	$5\frac{1}{8}$	$2\frac{1}{2}$	$7\frac{1}{2}$
3	3	$5\frac{7}{8}$	3	$8\frac{1}{4}$
4	4	$7\frac{1}{8}$	4	10
5	5	$8\frac{1}{2}$	5	11
6	6	$9\frac{7}{8}$	6	$12\frac{1}{2}$
8	8	$12\frac{1}{8}$	8	15
10	10	$14\frac{1}{4}$	10	$17\frac{1}{2}$
12	12	$16\frac{5}{8}$	12	$20\frac{1}{2}$
14	$13\frac{1}{4}$	$19\frac{1}{8}$	$13\frac{1}{4}$	23
16	$15\frac{1}{4}$	$21\frac{1}{4}$	$15\frac{1}{4}$	$25\frac{1}{2}$
18	17	$23\frac{1}{2}$	17	28
20	19	$25\frac{3}{4}$	19	$30\frac{1}{2}$
24	23	$30\frac{1}{2}$	23	36

*Includes 250 lb. and 800 lb Cast Iron; 300 lb. and higher Steel flanges.

Types of Gaskets
and Guide to their Use

Gasket Type	Steam				Water, Gas and Air	
	LP LT	LP HT	HP LT	HP HT	LP LT	HP LT
Soft rubber sheets with or without cloth reinforcement	—	—	—	—	X	—
Hard rubber sheets with or without cloth reinforcement	—	—	—	—	X	—
Fiber sheets	—	—	—	—	X	X
Asbestos composition	X	X	X	—	X	X
Corrugated metal, asbestos inserted, and spiral-wound metal, asbestos filled	X	X	X	X	X	X
Corrugated metal jacket, asbestos filled	X	X	X	X	X	X
Flat metal jacket, asbestos filled	X	X	X	X	X	X
Corrugated metal: copper	X	X	—	—	X	X
monel, steel and iron	—	—	X	X	X	X
Solid metal: aluminum	—	—	—	—	—	—
copper	X	X	X	—	X	X
monel, steel and iron	X	X	X	X	X	X

Notes: LT (low temperature) range: up to 500 deg F
　　　HT (high temperature) range: 500 to 1200 deg F
　　　LP (low pressure) range: up to 600 lb per sq in
　　　HP (high pressure) range: 600 to 2500 lb per sq in

Types of Gaskets
and Guide to their Use

Gasket Type	Oil		Oil Vapor		Refrig- erants	
	LP LT	LP HT	HP LT	HP HT	LP LT	HP LT
Soft rubber sheets with or without cloth reinforcement	—	—	—	—	X	—
Hard rubber sheets with or without cloth reinforcement	—	—	—	—	X	—
Fiber sheets	X	—	X	—	X	X
Asbestos composition	X	X	X	—	X	X
Corrugated metal, asbestos inserted, and spiral-wound metal, asbestos filled	X	X	X	X	X	X
Corrugated metal jacket, asbestos filled	X	X	X	X	X	X
Flat metal jacket, asbestos filled	X	X	X	X	X	X
Corrugated metal: copper	X	X	—	—	X	—
monel, steel and iron	X	X	X	X	X	X
Solid metal: aluminum	—	—	—	—	—	—
copper	X	X	X	X	X	X
monel, steel and iron	X	X	X	X	X	X

Notes, continued:
   Rubber is not recommended for use above 300 deg F
   Fiber is not recommended for use above 500 deg F
   Asbestos composition is not recommended above 800 deg F

Laying Lengths of
Cast Brass Solder Joint Ends

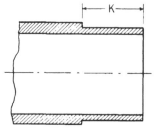

Male end

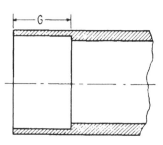

Female end

Nominal Size, Inches	Male End, K, Inches	Female End, G, Inches
¼	⅜	⁵⁄₁₆
⅜	⁷⁄₁₆	⅜
½	⁹⁄₁₆	½
¾	¹³⁄₁₆	¾
1	1	1⁵⁄₁₆
1¼	1¹⁄₁₆	1
1½	1³⁄₁₆	1⅛
2	1⁷⁄₁₆	1⅜
2½	1⁹⁄₁₆	1½
3	1¾	1¹¹⁄₁₆
3½	2	1¹⁵⁄₁₆
4	2¼	2³⁄₁₆
5	2¾	2¹¹⁄₁₆
6	3³⁄₁₆	3⅛
8	4¹⁄₁₆	4

K and G are given to nearest larger sixteenth.

Laying Lengths of
Cast Brass Solder 90° Elbows and Tees

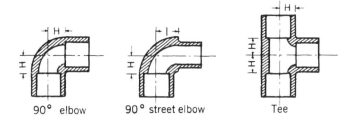

90° elbow   90° street elbow   Tee

Nominal Diameter, Inches	Dimension H, Inches	Dimension I, Inches
$\frac{1}{4}$	$\frac{1}{4}$	$\frac{3}{8}$
$\frac{3}{8}$	$\frac{5}{16}$	$\frac{7}{16}$
$\frac{1}{2}$	$\frac{7}{16}$	$\frac{9}{16}$
$\frac{3}{4}$	$\frac{9}{16}$	$\frac{11}{16}$
1	$\frac{3}{4}$	$\frac{7}{8}$
$1\frac{1}{4}$	$\frac{7}{8}$	1
$1\frac{1}{2}$	1	$1\frac{1}{8}$
2	$1\frac{1}{4}$	$1\frac{3}{8}$
$2\frac{1}{2}$	$1\frac{1}{2}$	$1\frac{5}{8}$
3	$1\frac{3}{4}$	$1\frac{7}{8}$
$3\frac{1}{2}$	2	—
4	$2\frac{1}{4}$	$2\frac{3}{8}$
5	$3\frac{1}{8}$	—
6	$3\frac{5}{8}$	—
8	$4\frac{7}{8}$	—

Laying Lengths of
Cast Brass Solder 45° Elbows

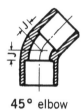 

45° elbow   45° street elbow

Nominal Diameter, Inches	Dimension J, Inches	Dimension Q, Inches
$\frac{1}{4}$	—	—
$\frac{3}{8}$	$\frac{3}{16}$	$\frac{5}{16}$
$\frac{1}{2}$	$\frac{3}{16}$	$\frac{5}{16}$
$\frac{3}{4}$	$\frac{1}{4}$	$\frac{3}{8}$
1	$\frac{5}{16}$	$\frac{7}{16}$
$1\frac{1}{4}$	$\frac{7}{16}$	$\frac{9}{16}$
$1\frac{1}{2}$	$\frac{1}{2}$	$\frac{5}{8}$
2	$\frac{9}{16}$	$\frac{3}{4}$
$2\frac{1}{2}$	$\frac{5}{8}$	—
3	$\frac{3}{4}$	—
$3\frac{1}{2}$	$\frac{7}{8}$	—
4	$\frac{15}{16}$	—
5	$1\frac{7}{16}$	—
6	$1\frac{5}{8}$	—
8	$2\frac{1}{8}$	—

Laying Lengths of
Cast Brass Solder Reducing 90° Elbows

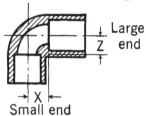

Large end

Z

→ X ←
Small end

Nominal Diameter, In.	Dimension X Inches	Dimension Z Inches
$\frac{3}{4}$ x $\frac{1}{2}$	$\frac{7}{16}$	$\frac{9}{16}$
1 x $\frac{3}{4}$	$\frac{5}{8}$	$\frac{3}{4}$
1 x $\frac{1}{2}$	$\frac{1}{2}$	$\frac{3}{4}$
$1\frac{1}{4}$x1	$\frac{3}{4}$	$\frac{7}{8}$
$1\frac{1}{2}$x$1\frac{1}{4}$	$\frac{7}{8}$	1
$1\frac{1}{2}$x $\frac{3}{4}$	$\frac{5}{8}$	1
2 x$1\frac{1}{2}$	1	$1\frac{1}{4}$
2 x1	$\frac{3}{4}$	$1\frac{1}{4}$
2 x $\frac{3}{4}$	$\frac{5}{8}$	$1\frac{1}{4}$
$2\frac{1}{2}$x2	$1\frac{1}{4}$	$1\frac{1}{2}$
$2\frac{1}{2}$x$1\frac{1}{2}$	1	$1\frac{1}{2}$
$2\frac{1}{2}$x1	$\frac{3}{4}$	$1\frac{1}{2}$
3 x$2\frac{1}{2}$	$1\frac{1}{2}$	$1\frac{3}{4}$
3 x$1\frac{1}{2}$	1	$1\frac{3}{4}$
3 x$1\frac{1}{4}$	$\frac{7}{8}$	$1\frac{3}{4}$
4 x3	$1\frac{3}{4}$	$2\frac{1}{4}$
4 x2	$1\frac{1}{4}$	$2\frac{1}{4}$
6 x4	$2\frac{5}{8}$	$3\frac{5}{8}$
8 x6	$3\frac{7}{8}$	$4\frac{7}{8}$

Laying Lengths of Cast Brass Solder Straight Couplings and Bushings

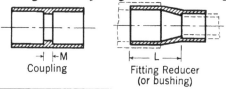

Coupling Fitting Reducer
(or bushing)

Straight Coupling		Reducer or Bushing	
Diameters, In.	M, In.	Diameters, In.	L, In.
$\frac{1}{4}$	$\frac{1}{16}$	$\frac{1}{2}$x $\frac{3}{8}$	$\frac{15}{16}$
$\frac{3}{8}$	$\frac{1}{16}$	$\frac{1}{2}$x $\frac{1}{4}$	$\frac{15}{16}$
$\frac{1}{2}$	$\frac{1}{8}$	$\frac{3}{4}$x $\frac{1}{2}$	$1\frac{3}{16}$
$\frac{3}{4}$	$\frac{1}{8}$	1 x $\frac{3}{4}$	$1\frac{1}{2}$
1	$\frac{1}{8}$	1 x $\frac{1}{2}$	$1\frac{1}{2}$
$1\frac{1}{4}$	$\frac{1}{8}$	$1\frac{1}{4}$x1	$1\frac{5}{8}$
$1\frac{1}{2}$	$\frac{1}{8}$	$1\frac{1}{4}$x $\frac{1}{2}$	$1\frac{5}{8}$
2	$\frac{3}{16}$	$1\frac{1}{2}$x$1\frac{1}{4}$	$1\frac{13}{16}$
$2\frac{1}{2}$	$\frac{3}{16}$	$1\frac{1}{2}$x $\frac{3}{4}$	$1\frac{13}{16}$
3	$\frac{3}{16}$	2 x$1\frac{1}{2}$	$2\frac{1}{8}$
$3\frac{1}{2}$	$\frac{1}{4}$	2 x$1\frac{1}{4}$	$2\frac{1}{8}$
4	$\frac{1}{4}$	2 x1	$2\frac{1}{8}$
5	$\frac{1}{4}$	$2\frac{1}{2}$x2	$2\frac{3}{8}$
6	$\frac{1}{4}$	$2\frac{1}{2}$x$1\frac{1}{2}$	$2\frac{3}{8}$
8	$\frac{5}{8}$	3 x$2\frac{1}{2}$	$2\frac{5}{8}$
—	—	3 x2	$2\frac{5}{8}$
—	—	3 x$1\frac{1}{2}$	$2\frac{5}{8}$
—	—	4 x3	$3\frac{7}{16}$
—	—	4 x$2\frac{1}{2}$	$3\frac{7}{16}$
—	—	4 x2	$3\frac{7}{16}$

Laying Lengths of
Cast Brass Solder Reducing Tees

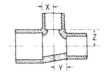

Nominal	Dimension, Inches		
Diameters, Inches	X	Y	Z
$\frac{3}{8}$x $\frac{3}{8}$x $\frac{1}{2}$	$\frac{7}{16}$	$\frac{7}{16}$	$\frac{3}{8}$
$\frac{3}{8}$x $\frac{3}{8}$x $\frac{1}{4}$	$\frac{1}{4}$	$\frac{1}{4}$	$\frac{5}{16}$
$\frac{1}{2}$x $\frac{1}{2}$x $\frac{3}{4}$	$\frac{9}{16}$	$\frac{9}{16}$	$\frac{7}{16}$
$\frac{1}{2}$x $\frac{1}{2}$x $\frac{3}{8}$	$\frac{3}{8}$	$\frac{3}{8}$	$\frac{7}{16}$
$\frac{1}{2}$x $\frac{1}{2}$x $\frac{1}{4}$	$\frac{5}{16}$	$\frac{5}{16}$	$\frac{7}{16}$
$\frac{1}{2}$x $\frac{3}{8}$x $\frac{1}{2}$	$\frac{7}{16}$	$\frac{7}{16}$	$\frac{7}{16}$
$\frac{1}{2}$x $\frac{3}{8}$x $\frac{3}{8}$	$\frac{3}{8}$	$\frac{3}{8}$	$\frac{7}{16}$
$\frac{3}{4}$x $\frac{3}{4}$x1	$\frac{3}{4}$	$\frac{3}{4}$	$\frac{5}{8}$
$\frac{3}{4}$x $\frac{3}{4}$x $\frac{1}{2}$	$\frac{7}{16}$	$\frac{7}{16}$	$\frac{9}{16}$
$\frac{3}{4}$x $\frac{3}{4}$x $\frac{3}{8}$	$\frac{3}{8}$	$\frac{3}{8}$	$\frac{9}{16}$
$\frac{3}{4}$x $\frac{1}{2}$x $\frac{3}{4}$	$\frac{9}{16}$	$\frac{9}{16}$	$\frac{9}{16}$
$\frac{3}{4}$x $\frac{1}{2}$x $\frac{1}{2}$	$\frac{7}{16}$	$\frac{7}{16}$	$\frac{9}{16}$
$\frac{3}{4}$x $\frac{1}{2}$x $\frac{3}{8}$	$\frac{3}{8}$	$\frac{3}{8}$	$\frac{9}{16}$
1 x1 x1$\frac{1}{2}$	1	1	$\frac{3}{4}$
1 x1 x1$\frac{1}{4}$	$\frac{7}{8}$	$\frac{7}{8}$	$\frac{3}{4}$
1 x1 x $\frac{3}{4}$	$\frac{5}{8}$	$\frac{5}{8}$	$\frac{3}{4}$
1 x1 x $\frac{1}{2}$	$\frac{1}{2}$	$\frac{1}{2}$	$\frac{3}{4}$
1 x1 x $\frac{3}{8}$	$\frac{7}{16}$	$\frac{7}{16}$	$\frac{3}{4}$
1 x $\frac{3}{4}$x1	$\frac{3}{4}$	$\frac{3}{4}$	$\frac{3}{4}$
1 x $\frac{3}{4}$x $\frac{3}{4}$	$\frac{5}{8}$	$\frac{5}{8}$	$\frac{3}{4}$

Laying Lengths of
Cast Brass Solder Reducing Tees

Nominal Diameters, Inches	Dimension, Inches		
	X	Y	Z
1 x ¾x ½	½	½	¾
1 x ½x1	¾	¾	¾
1 x ½x ¾	⅝	⅝	¾
1 x ½x ½	½	½	¾
1¼x1¼x2	1¼	1¼	⅞
1¼x1¼x1½	1	1	⅞
1¼x1¼x1	¾	¾	⅞
1¼x1¼x ¾	⅝	⅝	⅞
1¼x1¼x ½	½	½	⅞
1¼x1 x1¼	⅞	⅞	⅞
1¼x1 x1	¾	¾	⅞
1¼x1 x ¾	⅝	⅝	⅞
1¼x1 x ½	½	½	⅞
1¼x ¾x1¼	⅞	⅞	⅞
1¼x ¾x1	¾	¾	⅞
1¼x ¾x ¾	⅝	⅝	⅞
1¼x ½x1¼	⅞	⅞	⅞
1¼x ½x1	¾	¾	⅞
1½x1½x2½	1½	1½	1
1½x1½x2	1¼	1¼	1
1½x1½x1¼	⅞	⅞	1
1½x1½x1	¾	¾	1
1½x1½x ¾	⅝	⅝	1
1½x1½x ½	½	½	1

Laying Lengths of Cast Brass Solder Reducing Tees

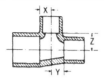

Nominal Diameters, Inches	Dimension, Inches		
	X	Y	Z
1½x1¼x1½	1	1	1
1½x1¼x1¼	⅞	⅞	1
1½x1¼x1	¾	¾	1
1½x1¼x ¾	⅝	⅝	1
1½x1¼x ½	½	½	1
1½x1 x1½	1	1	1
1½x1 x1¼	⅞	⅞	1
1½x1 x1	¾	¾	1
1½x ¾x1½	1	1	1
1½x ½x1½	1	1	1
2 x2 x4	2¼	2¼	1¼
2 x2 x3	1¾	1¾	1¼
2 x2 x2½	1½	1½	1¼
2 x2 x1½	1	1	1¼
2 x2 x1¼	⅞	⅞	1¼
2 x2 x1	¾	¾	1¼
2 x2 x ¾	⅝	⅝	1¼
2 x2 x ½	½	½	1¼
2 x1½x2	1¼	1¼	1¼
2 x1½x1½	1	1	1¼

Laying Lengths of
Cast Brass Solder Reducing Tees

Nominal Diameters, Inches	Dimension, Inches		
	X	Y	Z
2 x1½x1¼	⅞	⅞	1¼
2 x1½x1	¾	¾	1¼
2 x1½x ¾	⅝	⅝	1¼
2 x1½x ½	½	½	1¼
2 x1¼x2	1¼	1¼	1¼
2 x1¼x1½	1	1	1¼
2 x1¼x1¼	⅞	⅞	1¼
2 x1 x2	1¼	1¼	1¼
2 x ¾x2	1¼	1¼	1¼
2 x ½x2	1¼	1¼	1¼
2½x2½x4	2¼	2¼	1½
2½x2½x3	1¾	1¾	1½
2½x2½x2	1¼	1¼	1½
2½x2½x1½	1	1	1½
2½x2½x1¼	⅞	⅞	1½
2½x2½x1	¾	¾	1½
2½x2½x ¾	⅝	⅝	1½
2½x2½x ½	½	½	1½
2½x2 x2½	1½	1½	1½
2½x2 x2	1¼	1¼	1½
2½x2 x1½	1	1	1½
2½x2 x1¼	⅞	⅞	1½
2½x2 x1	¾	¾	1½
2½x2 x ¾	⅝	⅝	1½

Laying Lengths of
Cast Brass Solder Reducing Tees

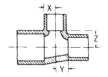

Nominal	Dimension, Inches		
Diameters, Inches	X	Y	Z
2½x2 x ½	½	½	1½
2½x1½x2½	1½	1½	1½
2½x1½x2	1¼	1¼	1½
2½x1½x1½	1	1	1½
2½x1¼x2½	1½	1½	1½
2½x1 x2½	1½	1½	1½
2½x ¾x2½	1½	1½	1½
2½x ½x2½	1½	1½	1½
3 x3 x4	2¼	2¼	1¾
3 x3 x2½	1½	1½	1¾
3 x3 x2	1¼	1¼	1¾
3 x3 x1½	1	1	1¾
3 x3 x1¼	⅞	⅞	1¾
3 x3 x1	¾	¾	1¾
3 x3 x ¾	⅝	⅝	1¾
3 x3 x ½	½	½	1¾
3 x2½x3	1¾	1¾	1¾
3 x2½x2½	1½	1½	1¾
3 x2½x2	1¼	1¼	1¾
3 x2½x1½	1	1	1¾

Laying Lengths of
Cast Brass Solder Reducing Tees

Nominal Diameters, Inches	Dimension, Inches		
	X	Y	Z
3 x2½x1¼	⅞	⅞	1¾
3 x2½x1	¾	¾	1¾
3 x2 x3	1¾	1¾	1¾
3 x2 x2½	1½	1½	1¾
3 x2 x2	1¼	1¼	1¾
3 x1½x3	1¾	1¾	1¾
3 x1¼x3	1¾	1¾	1¾
3 x1 x3	1¾	1¾	1¾
3½x3½x3	1¾	1¾	2
4 x4 x6	3⅝	3⅝	2⅝
4 x4 x3	1¾	1¾	2¼
4 x4 x2½	1½	1½	2¼
4 x4 x2	1¼	1¼	2¼
4 x4 x1½	1	1	2¼
4 x4 x1	¾	¾	2¼
4 x4 x ¾	⅝	⅝	2¼
4 x3 x4	2¼	2¼	2¼
4 x3 x3	1¾	1¾	2¼
4 x3 x2½	1½	1½	2¼
4 x3 x2	1¼	1¼	2¼
4 x2½x4	2¼	2¼	2¼
4 x2 x4	2¼	2¼	2¼
4 x2 x3	1¾	1¾	2¼
4 x2 x2	1¼	1¼	2¼

Laying Lengths of
Cast Brass Solder Reducing Tees

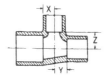

Nominal Diameters, Inches	Dimension, Inches		
	X	Y	Z
4 x1½x4	2¼	2¼	2¼
4 x1¼x4	2¼	2¼	2¼
4 x1 x4	2¼	2¼	2¼
5 x5 x4	2⅝	2⅝	3⅛
5 x4 x5	3⅛	3⅛	3⅛
6 x6 x8	4⅞	4⅞	3⅞
6 x6 x4	2⅝	2⅝	3⅝
6 x6 x3	2	2	3⅝
6 x6 x2½	1⅞	1⅞	3⅝
6 x6 x2	1⅝	1⅝	3⅝
6 x6 x1½	1⅜	1⅜	3⅝
6 x6 x1¼	1¼	1¼	3⅝
6 x6 x1	1⅛	1⅛	3⅝
6 x4 x6	3⅝	3⅝	3⅝
6 x4 x4	2⅝	2⅝	3⅝
6 x3 x6	3⅝	3⅝	3⅝
6 x2½x6	3⅝	3⅝	3⅝
6 x2 x6	3⅝	3⅝	3⅝
8 x8 x6	3⅞	3⅞	4⅞
8 x8 x4	2⅞	2⅞	4⅞

Laying Lengths of Cast Brass Solder Reducing and Eccentric Couplings

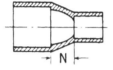 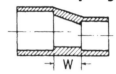

Reducing coupling   Eccentric coupling

Nominal Diameters, Inches	N, Inches	W, Inches
¾ x ½	$\frac{5}{16}$	$\frac{5}{8}$
1 x ¾	$\frac{3}{8}$	$\frac{11}{16}$
1¼x1	$\frac{3}{8}$	¾
1¼x ¾	$\frac{3}{8}$	¾
1½x1¼	$\frac{3}{8}$	$\frac{11}{16}$
1½x1	$\frac{3}{8}$	$\frac{11}{16}$
2 x1½	½	1⅛
2 x1¼	½	$\frac{15}{16}$
2 x1	½	—
2 x ¾	½	—
2½x2	$\frac{9}{16}$	$1\frac{3}{16}$
2½x1½	$\frac{9}{16}$	—
2½x1	$\frac{5}{8}$	—
3 x2½	$\frac{5}{8}$	1¼
3 x2	$\frac{5}{8}$	$1\frac{5}{16}$
4 x3	$\frac{11}{16}$	2
4 x2½	1⅛	—
4 x2	$1\frac{3}{16}$	—
6 x4	$1\frac{5}{16}$	—
8 x6	1⅜	—

Laying Lengths of
Cast Brass Elbows and Tees
with Internal Pipe Thread Ends

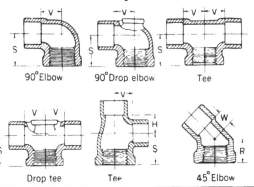

90°Elbow   90°Drop elbow   Tee

Drop tee   Tee   45°Elbow

Nominal Diameter, Inches	Dimension, Inches		
	S	V	H
$\frac{1}{4}$	$\frac{9}{16}$	$\frac{3}{8}$	—
$\frac{3}{8}$	$\frac{11}{16}$	$\frac{7}{16}$	$\frac{5}{16}$
$\frac{1}{2}$	$\frac{7}{8}$	$\frac{9}{16}$	$\frac{7}{16}$
$\frac{3}{4}$	1	$\frac{11}{16}$	$\frac{9}{16}$
1	$1\frac{1}{4}$	$\frac{7}{8}$	$\frac{3}{4}$
$1\frac{1}{4}$	$1\frac{1}{2}$	1	$\frac{7}{8}$
$1\frac{1}{2}$	$1\frac{5}{8}$	$1\frac{1}{8}$	1
2	$1\frac{15}{16}$	$1\frac{3}{8}$	$1\frac{1}{4}$
$2\frac{1}{2}$	$2\frac{1}{2}$	$1\frac{5}{8}$	—
3	$2\frac{13}{16}$	$1\frac{15}{16}$	—
4	$3\frac{7}{16}$	$2\frac{7}{16}$	—
6	$4\frac{7}{8}$	$3\frac{7}{8}$	—

Note: Dimension W is $\frac{3}{16}$, $\frac{3}{16}$, $\frac{1}{4}$ and $\frac{5}{16}$ and Dimension R is $\frac{11}{16}$, $\frac{15}{16}$, 1 and $1\frac{3}{16}$ for $\frac{3}{8}$, $\frac{1}{2}$, $\frac{3}{4}$ and 1 inch 45° elbows, respectively.

Laying Lengths of Cast Brass Solder Elbows and Tees with External Pipe Threads

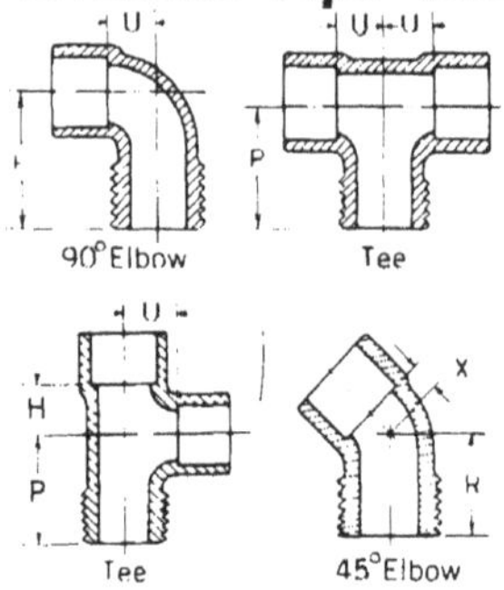

Cast-brass solder-joint fittings which have screwed ends shall have a right-hand thread conforming to the American Standard Taper Pipe Thread. The thread shall be concentric with the axis of the fittings.

Internal threads shall be chamfered approximately to the major diameter of the thread at the face of the fitting.

Nominal Diameter, Inches	Dimension, Inches		
	P	U	H
1/4	15/16	1/4	—
3/8	1 1/16	5/16	5/16
1/2	1 5/16	7/16	7/16
3/4	1 1/2	9/16	9/16
1	1 13/16	3/4	3/4
1 1/4	2	7/8	7/8
1 1/2	2 3/16	1	1
2	2 5/8	1 1/4	1 1/4

Note: Dimension X is 3/16, 3/16, 1/4 and 5/16 and Dimension R is 13/16, 1, 1 3/16 and 1 5/16 for 3/8, 1/2, 3/4 and 1 inch 45° elbows, respectively.

Laying Lengths of
Cast Brass Solder Joint Adapters

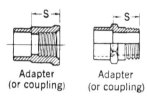

Adapter
(or coupling)

Adapter
(or coupling)

Fitting adapter
(or fitting coupling)

Fitting adapter
(or fitting coupling)

Diameter, Inches		Dimension, Inches		
Solder Joint	Pipe Thread	S	M	T
$\frac{1}{4}$	$\frac{1}{4}$	$\frac{5}{8}$	—	—
$\frac{3}{8}$	$\frac{1}{2}$	$\frac{3}{4}$	—	—
$\frac{3}{8}$	$\frac{3}{8}$	$\frac{5}{8}$	$\frac{5}{8}$	1
$\frac{1}{2}$	1	1	—	—
$\frac{1}{2}$	$\frac{3}{4}$	$\frac{7}{8}$	$\frac{7}{8}$	$1\frac{7}{16}$
$\frac{1}{2}$	$\frac{1}{2}$	$\frac{3}{4}$	$\frac{3}{4}$	$1\frac{1}{4}$
$\frac{1}{2}$	$\frac{3}{8}$	$\frac{5}{8}$	—	1
$\frac{3}{4}$	1	1	1	—.
$\frac{3}{4}$	$\frac{3}{4}$	$\frac{7}{8}$	$\frac{7}{8}$	$1\frac{7}{16}$
$\frac{3}{4}$	$\frac{1}{2}$	$\frac{3}{4}$	$\frac{3}{4}$	$1\frac{1}{4}$
1	$1\frac{1}{4}$	$1\frac{1}{16}$	$1\frac{1}{8}$	—
1	1	1	1	$1\frac{9}{16}$
1	$\frac{3}{4}$	$\frac{7}{8}$	—	$1\frac{1}{2}$
$1\frac{1}{4}$	2	$1\frac{1}{8}$	—	—

Laying Lengths of
Cast Brass Solder Joint Fitting Adapters

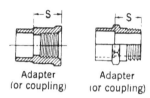

Adapter   Adapter
(or coupling)   (or coupling)

Fitting adapter   Fitting adapter
(or fitting coupling)   (or fitting coupling)

Diameter, Inches		Dimension, Inches		
Solder Joint	Pipe Thread	S	M	T
$1\frac{1}{4}$	$1\frac{1}{2}$	$1\frac{1}{16}$	—	$1\frac{3}{4}$
$1\frac{1}{4}$	$1\frac{1}{4}$	$1\frac{1}{16}$	$1\frac{1}{8}$	$1\frac{5}{8}$
$1\frac{1}{4}$	1	$1\frac{1}{16}$	—	$1\frac{5}{8}$
$1\frac{1}{2}$	2	$1\frac{1}{8}$	—	—
$1\frac{1}{2}$	$1\frac{1}{2}$	$1\frac{1}{16}$	$1\frac{3}{16}$	$1\frac{3}{4}$
$1\frac{1}{2}$	$1\frac{1}{4}$	$1\frac{1}{16}$	—	—
$1\frac{1}{2}$	1	1	—	—
2	2	$1\frac{1}{8}$	$1\frac{5}{16}$	$1\frac{7}{8}$
2	$1\frac{1}{2}$	$1\frac{1}{8}$	—	—
$2\frac{1}{2}$	$2\frac{1}{2}$	$1\frac{3}{8}$	$1\frac{11}{16}$	$2\frac{3}{8}$
3	3	$1\frac{1}{2}$	$1\frac{13}{16}$	$2\frac{9}{16}$
4	4	$1\frac{11}{16}$	$2\frac{1}{16}$	$2\frac{3}{4}$
6	6	2	$2\frac{9}{16}$	$5\frac{9}{16}$
8	8	$2\frac{1}{4}$	—	—

Soldering and Brazing

Solders. — Solders for joining metallic surfaces or edges are almost always composed of an alloy of two or more metals. The solder used must have a lower melting point than the metals to be joined by it, but the fusing point should approach, as nearly as possible, that of the metals to be joined so that a more tenacious joint is effected. Solders may be divided into two general classes, hard and soft. The former fuses at a red heat; the latter, at a comparatively low temperature.

Soft Solders. — Soft solders consist chiefly of lead and tin, although other metals are occasionally added to lower the melting point. Lead-tin alloys melt at a lower temperature, with an increase in the percentage of tin, up to a certain point, but when the tin exceeds 67 per cent, the melting point rises gradually to the melting point of tin. Soft solders are termed "common," "medium" and "fine," according to the tin content, those containing the most lead being the cheapest and having the highest melting temperatures.

Fine solder is largely used for soldering britannia metal, brass and tin-plate articles. It is also used for soldering cast iron, steel, copper and many alloys. The soft solder called "common" is used by plumbers for ordinary work; this solder contains two parts of lead to one part of tin. The best soft solders are made from pure lead and pure tin. Antimony is an objectionable impurity as it renders the solder less fluid when melted and tends to prevent perfect adhesion of the surfaces. Zinc also has an injurious effect on soft solder, causing it to flow sluggishly. Aluminum acts in a similar way. A small percentage of phosphorus renders soft solder very "lively"; that is, the solder has a tendency to run freely. Too much phosphorus is injurious, and if added to thin the solder it should be in the form of phosphor-tin.

Hard Soldering and Brazing. — Hard solder is used for joining such metals as copper, silver and gold, and alloys such as brass, German silver, gun metal, etc., which require a strong joint and often a solder the color

Melting Temperatures of Lead-tin Alloys

Percentage		Melting Temp., Deg. F.	Percentage		Melting Temp., Deg. F.
Tin	Lead		Tin	Lead	
0	100	618.8	60	40	368.6
10	90	577.4	66	34	356.0
20	80	532.4	70	30	365.0
30	70	491.0	80	20	388.4
40	60	446.0	90	10	419.0
50	50	401.0	100	0	450.0

Melting Temperatures of Copper-zinc Alloys

Percentage		Melting Temp., Deg. F.	Percentage		Melting Temp., Deg. F.	Percentage		Melting Temp., Deg. F.
Copper	Zinc		Copper	Zinc		Copper	Zinc	
100	0	1980	71	29.	1746	41	59	1544
96	4	1967	66.4	33.6	1684	35	65	1501
86	14	1890	63	37	1666	33	67	1477
80	20	1846	60	40	1634	29	71	1467
76	24	1796	50	50	1616	24	76	1364
72	28	1756	48	52	1598	20	80	1301

of which is near that of the metal to be joined. The hard soldering of copper, iron, brass, etc., is generally known as brazing, and the solder as spelter. The operations of hard soldering and brazing are identical, and the two terms are often used interchangeably. According to common usage, however, there is the following distinction. Brazing is generally understood to mean the joining of metals by a film of brass, whereas hard soldering (which is the term used by jewelers) ordinarily means that "silver solder" is used as the uniting medium. For hard soldering or brazing, a red heat is necessary, and borax is used as a flux to protect the metal from oxidation, and to dissolve the oxides formed. Heating cannot be done with a soldering iron, but should be effected by a blowpipe, blowtorch, gas forge or a coke or charcoal fire.

As a greater degree of heat is required to melt spelter than soft solder, brazed work will withstand more heat without breaking or weakening than parts which are soldered. The chief advantage of a brazed joint, however, lies in its superior strength. Before work is assembled for brazing, it should be carefully cleaned; the parts are then fastened together in the position they are to occupy when joined.

Fluxes for Soldering. — As two pieces to be soldered must be thoroughly alloyed with the material used as a solder, the temperature must be raised and maintained at such a point that inter-penetration can take place completely. It is necessary that the surfaces to be joined be perfectly clean, and means must be provided to prevent oxidation during soldering, oxides tending to prevent interfusion. This is accomplished by using a coating of some substance that melts at the fusing temperature of the solder, and thus excludes the air.

Alloys for Brazing Solders. — The alloys or spelters used for brazing are composed of copper-zinc alloys. The melting point of these alloys depends upon the percentage of zinc. As the proportion of zinc increases, the melting point is lowered. The fusing point of the spelter should be as close as possible to that of the article to be brazed, as a more tenacious joint is thereby secured. An easily fusible spelter may be made of two parts zinc to one part copper, but the joint will be weaker than when an alloy more difficult to fuse is employed. A spelter that is readily fused may be made of 44 per cent copper, 50 per cent zinc, 4 per cent tin and 2 per cent lead. Alloys containing much lead should be avoided, since lead does not transfuse with brass and thus decreases the strength of the joint. A hard solder for the richer alloys of copper and zinc may be produced from 53 parts copper and 47 parts zinc. Copper and iron have a much higher melting point than brass, thus allowing the use of a richer copper alloy.

Composition of Brazing Alloys

Percentage				Characteristics	Color
Copper	Zinc	Tin	Lead		
58	42	—	—	Very strong	Reddish-yellow
53	47	—	—	Strong	Reddish-yellow
48	52	—	—	Medium	Reddish-yellow
54.5	43.5	1.5	0.5	Medium	Reddish-yellow
34	66	—	—	Easily fusible	White
44	50	4	2	Easily fusible	Gray
55	26	15	4	White solder	White

Soft and Hard Solders for Various Metals

Metal to be Soldered	Flux	Soft Solder	
		Tin	Lead
Brass	Chloride of zinc, rosin, or chloride of ammonia	66	34
Gun metal		63	37
Copper		60	40
Lead	Tallow or rosin	33	67
Block tin	Chloride of zinc	99	1
Tinned steel	Chloride of zinc or rosin	64	36
Galvanized steel	Hydrochloric acid	58	42
Zinc	Hydrochloric acid	55	45
Iron and steel	Chloride of ammonia	50	50

Metal to be Soldered	Flux	Hard Solder	
		Copper	Zinc
Brass, soft	Borax	22	78
Brass, hard	Borax	45	55
Copper	Borax	50	50
Cast iron	Cuprous oxide	55	45
Iron and steel	Borax	64	36

Sweating. — When parts are soldered together by heating them sufficiently to melt the solder, instead of using a soldering iron, the operation is often known as sweating. The finished surfaces forming the joint are first tinned or covered with solder. This is done by heating enough to melt the solder, then applying a flux (such as sal-ammoniac), and finally the solder.

Dimensions of U-Bolts for Pipe Hangers

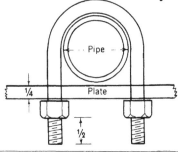

Nominal Pipe Size, Inches	Bolt Diameter, Inches			
	1/4	3/8	1/2	5/8
	Length of Bolt, Inches			
1/2	5 1/4	6 1/8	—	—
3/4	5 3/4	6 3/4	—	—
1	6 1/4	7 3/8	—	—
1 1/4	7 1/8	8	—	—
1 1/2	7 7/8	8 3/4	—	—
2	8 7/8	9 1/2	—	—
2 1/2	9 1/4	9 1/4	11 1/8	11 1/8
3	11 3/4	11 3/4	12 1/4	12 1/4
3 1/2	13	13	14 1/4	14 1/4
4	14 3/8	14 3/8	15 3/8	15 3/8

Nominal Pipe, Size, Inches	Bolt Diameter, Inches		
	3/4	7/8	1
	Length of Bolt, Inches		
5	19	19	19 3/4
6	21 5/8	21 5/8	22 3/8
8	26 7/8	26 7/8	27 7/8
10	32 1/2	32 1/2	37 1/2

Spacing of Pipe Supports
for Lines Carrying Water†
(Standard Weight Pipe)

Nominal Dia., In.	Horizontal Runs			Sloping Runs‡	
	Atmospheric Temp.	200° Temp.	400° Temp.	Grade 1 Inch in 10 Ft.	Grade 1 Inch in 20 Ft.
	Spacing of Supports in Feet				
1	16	11	—	11	—
1¼	18	12	11	13	11
1½	19	13	12	15	12
2	22	15	13	18	14
2½	24	16	15	21	16
3	27	18	16	24	19
3½	28	20	18	27	21
4	30	21	18	30	23
5	33	23	20	35	27
6	36	25	22	39	31
8	40	27	25	48	37
10	43	30	27	50	43
12	46	31	29	50	49
14	48	32	30	50	50
16	50	33	31	50	50
18	50	34	32	50	50
20	50	35	33	50	50
24	50	36	34	50	50

† This table, based on average conditions, does not include allowance for flanges, fittings and valves, but does include allowance for insulation. Heavy fittings such as valves should always be located close to supports. Wider spans may be used where data are calculated. Data under "Horizontal Runs" are based on stress, those under "Sloping Runs" are based on gradient.

‡ Spacing should not exceed that given for appropriate temperature in columns to the left.

Data courtesy Tube Turns, Inc.

Spacing of Pipe Supports
for Lines Carrying Gas or Steam†
(Standard Weight Pipe)

Nominal Dia., In.	Horizontal Runs			Sloping Runs‡	
	Atmos-pheric Temp.	200° Temp.	800° Temp.	Grade 1 Inch in 10 Ft.	Grade 1 Inch in 20 Ft.
	Spacing of Supports in Feet				
1	15	12	—	11	—
1¼	18	14	—	13	11
1½	20	15	—	15	12
2	22	17	11	18	14
2½	25	19	12	20	15
3	28	21	14	24	19
3½	31	23	15	26	21
4	33	24	16	29	23
5	37	27	18	34	27
6	40	30	20	38	31
8	46	34	24	47	37
10	50	38	27	50	43
12	50	42	30	50	49
14	50	44	31	50	50
16	50	47	34	50	50
18	50	50	36	50	50
20	50	50	38	50	50
24	50	50	41	50	50

† This table, based on average conditions, does not include allowance for flanges, fittings and valves, but does include allowance for insulation. Heavy fittings such as valves should always be located close to supports. Wider spans may be used where data are calculated. Data under "Horizontal Runs" are based on stress, those under "Sloping Runs" are based on gradient.

‡ Spacing should not exceed that given for appropriate temperature in columns to the left.

Data courtesy Tube Turns, Inc.

Weight of Water per Foot
of Pipe and Tube

Nominal Diameter, Inches	Type of Pipe or Tube				
	Sched. 40 Steel or Iron	Sched. 80 Steel or Iron	Type K Copper	Type L Copper	Type M Copper
	Pounds of Water per Lineal Foot				
⅛	0.028	0.0158	.0119	.0138	.0138
¼	0.045	0.0310	.0325	.0338	.0363
⅜	0.083	0.0610	.0550	.0625	.0688
½	0.132	0.1020	.0944	.1013	.1100
¾	0.232	0.2130	.1894	.2100	.2238
1	0.375	0.3120	.3375	.3581	.3794
1¼	0.649	0.5550	.5281	.5450	.5688
1½	0.882	0.7650	1.2238	.7731	.8331
2	1.454	1.2800	1.3075	1.3419	1.3456
2½	2.073	1.8300	2.0200	2.0700	2.1225
3	3.201	2.8700	2.8794	2.9550	3.0294
3½	4.287	3.7200	———	———	———
4	5.516	4.9700	5.0706	5.1956	5.2300
5	8.674	7.9400	7.8669	8.0988	8.8031
6	12.520	11.3000	11.2300	11.6388	11.7838
8	21.680	19.8000	19.5906	20.3338	20.6525
10	34.160	31.1300	30.4188	31.5631	32.0663
12	48.500	44.0400	43.6113	45.5700	45.9806
14	58.640	53.1800	———	———	———
16	76.580	69.7300	———	———	———
18	96.930	88.5000	———	———	———
20	120.460	109.5100	———	———	———
24	174.230	158.2600	———	———	———

Gallons of Water per Foot
of Pipe and Tube

Nominal Diameter, Inches	Type of Pipe or Tube				
	Sched. 40 Steel or Iron	Sched. 80 Steel or Iron	Type K Copper	Type L Copper	Type M Copper
	Gallons of Water per Lineal Foot				
⅛	0.003	0.0019	.00142	.0016	.00165
¼	0.005	0.0037	.00389	.0040	.00435
⅜	0.010	0.0073	.00658	.0075	.00823
½	0.016	0.0122	.01129	.0121	.01316
¾	0.028	0.0255	.02664	.0251	.02678
1	0.045	0.0374	.04039	.0429	.04540
1¼	0.077	0.0666	.06321	.0652	.06807
1½	0.106	0.0918	.14646	.0925	.09971
2	0.174	0.1535	.15648	.1606	.16463
2½	0.248	0.2200	.24175	.2478	.25402
3	0.383	0.3440	.34460	.3537	.36256
3½	0.513	0.4580	————	————	————
4	0.660	0.5970	.60682	.6218	.62593
5	1.040	0.9470	.94151	.9693	.98175
6	1.500	1.3550	1.3440	1.393	1.4103
8	2.600	2.3800	2.3446	2.434	1.4717
10	4.100	4.1650	3.4405	3.777	3.838
12	5.870	5.2800	5.2194	5.454	5.503
14	7.030	6.3800	————	————	————
16	9.180	8.3600	————	————	————
18	11.120	10.6100	————	————	————
20	14.400	13.1300	————	————	————
24	20.900	19.0000	————	————	————

Expansion of Pipe per 100 Feet

The table below is used to determine the expansion of pipe from one temperature to a higher one.

EXAMPLE

A 740-foot steel pipe line was installed during 60° weather. What is the computed expansion of the line when 400° steam is turned on in the line?

SOLUTION

Under "Steel," opposite 400°, find 3.245; opposite 60° find 0.449. Subtracting gives 2.796 in. expansion per 100 feet. For 740 feet the expansion is 7.4 × 2.796 = 20.69 inches.

Temp., F	Material		
	Steel	Wrought Iron	Copper
0	0	0	0
20	0.149	0.156	0.222
40	0.299	0.313	0.444
60	0.449	0.470	0.668
80	0.601	0.629	0.893
100	0.755	0.791	1.119
120	0.909	0.952	1.346
140	1.066	1.115	1.575
160	1.224	1.281	1.805
180	1.384	1.447	2.035
200	1.545	1.616	2.268
220	1.708	1.786	2.501
240	1.872	1.957	2.736
260	2.038	2.130	2.971
280	2.207	2.305	3.208
300	2.376	2.481	3.446
320	2.547	2.659	3.685
340	2.718	2.838	3.926
360	2.892	3.017	4.167
380	3.069	3.199	4.411
400	3.245	3.383	4.653
500	4.148	4.327	5.892
600	5.096	5.309	7.160
700	6.083	6.351	8.460
800	7.102	7.384	9.783
900	8.172	8.489	11.144
1000	9.275	9.627	12.532
1100	10.042	10.804	13.950
1200	11.598	12.020	15.397

Contents of Cylindrical Tanks
in U. S. Gallons

Length of Tank, Feet	Diameter of Tank, Feet				
	5	6	7	8	9
	Contents of Tank, U. S. Gallons				
5	734	1058	1439	1880	2379
6	881	1269	1727	2256	2855
7	1028	1481	2015	2632	3331
8	1175	1692	2303	3008	3807
9	1322	1904	2591	3384	4283
10	1469	2115	2879	3760	4759
11	1616	2327	3167	4136	5235
12	1763	2538	3455	4512	5711
13	1909	2750	3742	4888	6187
14	2056	2961	4030	5264	6662
15	2203	3173	4318	5640	7138
16	2350	3384	4606	6016	7614
17	2497	3596	4894	6392	8090
18	2644	3807	5182	6768	8566
19	2791	4019	5480	7144	9042
20	2938	4230	5758	7520	9518

Contents of Cylindrical Tanks
in U. S. Gallons

Length of Tank, Feet	Diameter of Tank, Feet				
	10	11	12	14	16
	Contents of Tank, U. S. Gallons				
5	2938	3555	4230	5758	7521
6	3525	4265	5076	6909	9025
7	4113	4976	5922	8061	10529
8	4700	5687	6768	9212	12033
9	5288	6398	7614	10364	13537
10	5875	7109	8460	11515	15041
11	6463	7820	9306	12667	16545
12	7050	8531	10152	13818	18049
13	7638	9242	10998	14970	19553
14	8225	9953	11844	16121	21057
15	8813	10664	12690	17273	22562
16	9400	11374	13536	18424	24066
17	9988	12085	14383	19576	25570
18	10575	12796	15229	20727	27074
19	11163	13507	16075	21879	28578
20	11750	14218	16921	23030	30082

Contents of Cylindrical Tanks
in U. S. Gallons

Length of Tank, Feet	Diameter of Tank, Feet				
	18	20	22	24	25
	Contents of Tank, U. S. Gallons				
5	9518	11751	14218	16921	18360
6	11422	14101	17062	20305	22032
7	13325	16451	19905	23689	25704
8	15229	18801	22749	27073	29376
9	17132	21151	25592	30457	33048
10	19036	23501	28436	33841	36720
11	20940	25851	31280	37225	40392
12	22843	28201	34123	40609	44064
13	24747	30551	36967	43993	47736
14	26650	32901	39810	47377	51408
15	28554	35252	42654	50762	55080
16	30458	37602	45498	54146	58752
17	32361	39952	48341	57530	62424
18	34265	42302	51185	60914	66096
19	36168	44652	54028	64298	69768
20	38072	47002	56872	67682	73440

Identification of Piping Systems

Schemes for the identification of piping systems have been developed in the past by a large number of industrial plants and organizations. The schemes arrived at may have given complete satisfaction to these using them but they have suffered from a lack of uniformity. Considerable confusion as well as accidents have occurred to those who change employment from one plant to another and to outside agencies, such as municipal fire departments, when called in to assist.

In order to promote greater safety, and lessen the chances of error, confusion, or inaction, especially in times of emergency, a uniform code for identification of piping has been established by the American Standards Association and published by The American Society of Mechanical Engineers. The standard is based on primary identification of the contents of a piping system by stenciled legend and secondary identification through the use of color. It is urged that industry and organizations do not use color as a means of specifying the type of material contained in a piping system unless its use is in conformity with provisions of the standard and supplementary to the use of legends.

Any material transported in a piping system will fall into one of four main classifications.

Fire Protection Materials and Equipment. This classification includes sprinkler systems and other fire fighting or fire protection equipment. The identification for this group may also be used to locate such equipment as alarm boxes, extinguishers, fire doors, hose connections, and hydrants.

Dangerous Materials. This group includes materials which are hazardous to life or property because they are easily ignited, corrosive at high temperatures and pressures, productive of poisonous gases, or are in themselves poisonous.

Safe Materials. This group includes materials involving little or no hazard to life or property in their handling. Classification embraces materials at low pressures and temperatures which are not poisonous and will not produce fire or explosion.

Protective Materials. This group includes materials which are piped through plants for the express purpose of being available to prevent or minimize the hazard of the dangerous materials previously mentioned.

(There formerly was a fifth classification, Extra Valuable Materials, identified by the color purple).

Method of Identification. Positive identification of contents of a piping system shall be by lettered legend giving the name of the material in full or abbreviated form. Arrows may be used to indicate the direction of flow. Where it is desirable or necessary to give supplementary information such as hazard or use of the piping system contents, this may be done by additional legend, by color applied to the entire piping system, or by colored bands.

Color Identification of Classifications

Classification	Predominant Color of System	Color of Letters for Legends
F — Fire protection materials and equipment	Red	White
D — Dangerous materials	Yellow (or orange)	Black
S — Safe materials	Green (or the achromatic colors: white, black, gray, or aluminum)	Black
P — Protective materials	Bright blue	White

Size of Color Bands and Legend Letters

Outside Diameter of Pipe or Covering Inches	Width of Color Bands Inches	Height of Legend Letters Inches
¾ to 1¼	8	½
1½ to 2	8	¾
2½ to 6	12	1¼
8 to 10	24	2½
Over 10	32	3½

Identifying Colors of Typical Materials Transported in Piping Systems

Material	Color	Material	Color
Acetic Acid	Yellow	Chloroform	Yellow
Acetone	Yellow	Circulating water	Green
Acetylene gas	Yellow	City gas	Yellow
Acid	Yellow	City water	Green
Air	Green	Coal gas	Yellow
Alcohol	Yellow	Cold water	Green
Alum	Green	Compressed Air	Green
Ammonia	Yellow	Condensate	Yellow
Ammonium nitrate	Yellow	Cooling water	Green
Amyl acetate	Yellow	Cottonseed oil	Yellow
Antidote gas	Blue	Cutting oil	Green
Argon	Green	Diesel oil	Yellow
Benzol	Yellow	Distilled water	Green
Bisulphite liquor	Yellow	Drain oil	Yellow
Blau gas	Yellow	Drain water	Green
Bleach liquor	Yellow	Drinking water	Green
Blow-off water	Yellow	Dye	Yellow
Boiler feed water	Yellow	Ethane dye	Yellow
Brine	Green	Exhaust air	Green
Burner gas	Yellow	Exhaust gas	Yellow
Butane	Yellow	Exhaust system	Yellow
Butyl alcohol	Yellow	Filtered water	Blue
Calcium chloride	Blue	Fire protection water	Red
Carbon bisulphide	Yellow	Flue gas	Yellow
Carbon dioxide	Yellow	Foamite	Red
Carbon monoxide	Yellow	Formalin	Yellow
Carbonated water	Green	Freon	Green
Caustic soda	Yellow	Fresh water	Green
Chlorine	Yellow	Fuel gas	Yellow
Chlorine gas	Yellow	Fuel oil	Yellow

Identifying Colors of Typical Materials Transported in Piping Systems

(Continued)

Material	Color	Material	Color
Gas	Yellow	Phenol	Yellow
Gasoline	Yellow	Process gas	Yellow
Glycerine	Green	Producer gas	Yellow
Heating Returns	Yellow	Propane gas	Yellow
Heating Steam	Yellow	Raw water	Green
Hot water	Yellow	Refrigerated water	Green
Hydrochloric acid	Yellow	River water	Green
Hydrogen	Yellow	Salt water	Green
Hydrogen peroxide	Yellow	Sanitary sewer	Green
Hydrogen sulphide	Yellow	Soda ash	Green
Instrument air	Green	Solvent	Yellow
Kerosene	Yellow	Soybean oil	Yellow
Lacquer	Yellow	Sprinkler, water	Red
Linseed oil	Yellow	Steam	Yellow
Lubricating oil	Yellow	Storm sewer	Green
Make-up water	Green	Sugar juice	Green
Mercury	Yellow	Sulphur chloride	Yellow
Methyl chloride	Yellow	Sulphur dioxide	Yellow
Mixed acid	Yellow	Sulphuric acid	Yellow
Mixed gas	Yellow	Tar	Yellow
Muriatic acid	Yellow	Toluene	Yellow
Naphtha	Yellow	Toluol	Yellow
Natural gas	Yellow	Trichloroethylene	Yellow
Nitric acid	Yellow	Turpentine	Yellow
Nitrogen	Green	Vapor	Yellow
Nitrogen oxide	Yellow	Varnish	Yellow
Oil	Yellow	Vegetable oil	Yellow
Oxygen	Yellow	Waste water	Green
Paint	Yellow	Water	Green
Peanut oil	Yellow	Water gas	Yellow

Gate Valves

Description: The working parts of gate valves include a solid or split wedge, or gate, which fits into the open passageway of the valve between machined seats; a threaded stem or spindle; a handwheel; and packing.

The bonnet may be of one piece construction screwed directly to the valve body, or consist of a union connection screwed to the body. Bonnet may also be bolted to the body, or be constructed as a yoke, exposing the stem or spindle.

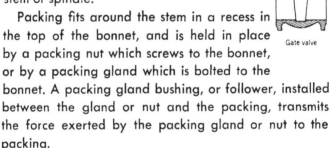

Gate valve

Packing fits around the stem in a recess in the top of the bonnet, and is held in place by a packing nut which screws to the bonnet, or by a packing gland which is bolted to the bonnet. A packing gland bushing, or follower, installed between the gland or nut and the packing, transmits the force exerted by the packing gland or nut to the packing.

Service Characteristics. Flow through gate valves is straight-way. They are best for lines where unrestricted flow is important — in pump lines, main supply lines, and for stop-valve service. They are suited for service in which valves are infrequently operated, with gate either wide open or fully closed. Gate valves are not considered suitable for throttling services (flow regulation).

The split or double disc type should be installed with stem vertical, handwheel up, as a precaution against possible jamming of the disc spreader mechanism. This type is good for non-condensing gas and liquid services at normal temperatures.

The solid wedge type can be installed in any position without danger of the disc jamming. It is recognized as best for steam service, and has highest resistance to pressure strains.

Globe Valves

Description. The working parts of globe valves consist of a disc which fits over a circular horizontal opening in the valve passageway into which a seat has been fitted or machined; a stem or spindle; a handwheel; and packing. The same variations that apply to the construction of bonnets, packing glands or nuts, and the types of end connections of gate valves also apply to globe valves.

Service Characteristics. Globe valves are well suited to flow regulation (throttling) by hand. Globe valve design causes a change in the direction of flow through valve body, with increased resistance to flow. On liquid lines, pump lines, etc., this may be objectionable. Good for frequency of operation, the valves have short stem travel which saves operator's time. Convenient and quick regrinding feature of globe valves makes them highly suitable for severe services requiring frequent repair.

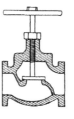

Globe valve

Installation of globe valves depends on type of fluid service to be regulated. If continuous flow is desired in the event that the disc becomes detached from the stem, valve should be installed in line so that line pressure is under disc. If flow stoppage is desired under this circumstance, valve should be installed with line pressure on top of disc.

Rising stem, inside screw type is simplest and most common stem construction. Stem rises when valve is opened, and thus indicates position of disc.

Angle Valves. Angle valves have the same basic characteristics as globe valves, and are available in a similar range of constructions. Used when making a 90 deg turn in a line, an angle valve gives less restriction to flow than the elbow and globe valve which it displaces.

Check Valves

Description. Check valves are devices designed to allow a fluid to pass through in one direction only. There are two basic types — the swing check valve and the lift check valve.

Working parts of swing check valves consist of a hinged disc or clapper which is free to swing upon a hinge pin in only one direction, that being the direction of the flow of liquid in the pipe line. The pressure exerted by the liquid flowing through the valve lifts the clapper and holds it in an open position. When flow stops in that direction, the clapper falls back to its original position by gravity, thus preventing back flow.

Working parts of lift check valves consist of a valve disc so positioned that it is free to rise and fall in a vertical direction. Pressure exerted by the liquid flowing through the valve raises the disc and holds it in an open position. When flow stops in that direction, the disc drops back to a closed position by gravity, thus preventing flow. Any force or tendency to cause back flow only causes the closing action to become stronger in both types of check valves.

There are no exterior parts whatever on a check valve. Failure of the internal parts cannot be determined by inspection, except by disassembling the valve.

Service Characteristics. Flow through a swing check valve is straight-way, as in gate valves. Lift check design, as in globe valves, requires a change in direction of flow through the valve body. A safe rule for choosing check valves is to use swing checks in combination with gate valves; use lift checks with globe and angle valves.

Trouble Shooting Methods

Trouble and Cause	Remedy
Gate Valves	
Leakage around stem	
Loose packing gland or nut	Tighten packing gland or nut
Packing worn out	Replace packing
Bent or scored stem	Replace stem
Leakage around gasketed, screwed or soldered connections	
Loose connections	Tighten flange bolts or screwed connections, or resolder joints
Gasket failure	Replace gasket
Handwheel turns without stopping at extremes	
Broken stem	Replace stem
Valve does not close tightly	
Wedge damaged	Replace wedge
Seat damaged	Replace valve
Wedge-stem joint damaged	Replace stem and/or wedge
Deposits under wedge	"Wash" or clean

Trouble Shooting Methods, Continued

Trouble and Cause	Remedy
Globe Valves	
Leakage around stem	
Loose packing gland or nut	Tighten packing gland or nut
Packing worn out	Replace packing
Bent or scored stem	Replace stem
Leakage around gasketed, screwed or soldered connections Loose connection	Tighten flange bolts Tighten screwed connections or resolder joints
Gasket failure	Replace gasket
Valve does not close tightly	
Disc damaged	Replace disc
Seat damaged	Replace valve
Disc-stem joint damaged	Replace disc or stem
Deposits under disc	"Wash" or clean
Check Valves	
Leakage around gasketed, screwed or soldered connections Loose connection	Tighten flange bolts or screwed connections, or resolder joints
Gasket failure	Replace gasket

Plastic Pipe

Thermoplastics are a relatively new entrant to the field of materials manufactured for piping purposes. The field is a dynamic one; research and development are being done not only to improve characteristics and extend applications of existing types, but also to introduce new types of thermoplastics for piping.

Thermoplastics are, by definition, plastics which soften upon application of heat and reharden upon cooling. They can be softened and hardened repeatedly. In general, advantages claimed for all thermoplastics include resistance to chemical and electrolytic attack, light weight, ease of installation, and less pressure loss and turbulence.

It should be emphasized that plastic pipe is not to be considered a substitute for other materials but rather an entirely new type of piping with its own special capabilities and limitations. At present, there are four principal thermoplastics with a wide range of practical use in piping applications, and two of these have, respectively, two and three commercial grades. Each of these seven, then, while having some overlapping applications, has individual characteristics and recommended utilization.

There are other types and forms of plastics used for piping, but due to their rather specialized fields of application, they will not be discussed.

Polyvinyl Chloride (PVC)

For general excellence of properties, especially chemical resistance, tensile strength, and temperature rating, polyvinyl chloride is seen, at present, as having the greatest all-around application of all the plastic pipe materials. It is available not only as pipe, but also in complete and standardized lines of threaded and socket-type fittings, threaded and socket-type flanges, valves, and pumps. For corrosive atmospheres, even PVC bolts and nuts are available for flanged connections.

Joining Methods

There are three recommended methods of joining PVC pipe:

1. *Solvent Welding.* This is accomplished with a solvent cement and paint brush. After pipe end is cut square, using a hand saw and miter box or a power saw, cement is brushed on end of pipe and socket-end of fitting or flange (after thorough cleaning with acetone or carbon tetrachloride). Pipe is then pressed firmly into fitting and turned slightly for even distribution of cement. Handling strength is developed after 30 min; full strength of the solvent weld is not reached for 24 hr or more.

2. *Threading.* This method is not recommended for Schedules A and 40, or in any case where operating temperature will exceed 120 deg F. Using standard hand or machine pipe-threading tools with sharp dies, Schedules 80 and 120 can be easily threaded without the use of cutting lubricants. Pads should be placed in pipe vise jaws to prevent scoring of pipe. A tapered plug should be inserted in pipe to insure thread uniformity. In threaded assembly, screwed fittings should be started carefully, hand-tightened, and then further tightened with a strap wrench. Standard pipe wrenches cannot be used since they deform and scar the pipe, thus weakening it.

3. *Hot Welding.* Hot air fusion welding of PVC pipe can be done effectively after instruction and practice in the technique. When properly made, welds have average tensile strength of 80% to 90% of the PVC material. Hot air welding equipment and filler rods designed for PVC welding are available from several manufacturers.

Pipe Bending

All PVC pipe bending must be done hot, with the section to be bent being heated to 250-275 deg F. Pipe should be uniformly heated in a circulated hot air oven, by immersion in heated oil, by hot air torch, or by other heating devices.

By bending pipe around a regular pipe bending form grooved to the pipe diameter, it is possible to obtain a bend radius not less than five times the diameter of the pipe without evidence of flattening. For sizes of 2 inches and above, or for severe bends, the pipe should be filled with hot dry sand or a coil spring before bending to minimize flattening.

Because of the springback characteristic of PVC, the pipe should be bent slightly beyond the desired shape, and quickly cooled with compressed air or water while held in the bent position.

Pipe Supporting

Roll, ring, angle, or spring hangers may be used to support PVC pipe. Most favorable is the clevis or saddle-type hanger, except where thrust along axis of pipe must be controlled as in thermal expansion. For firm anchoring, metal compression hangers are satisfactory only when padded with a compressible insert band.

Pipe lines should have additional support to that recommended in the table at fittings and flanges. Valves should be supported independently, and braced to resist twisting during opening and closing.

Continuous supports are advisable for pipe carrying hazardous fluids, and for lines operating at the upper temperature limits of the PVC pipe: 160 deg for Type I—Normal Impact, and 150 deg for Type II—High Impact.

Buried Lines

For all buried lines, cemented (solvent welded) joints are recommended. An additional one-bead seal weld (hot air welded) will give added assurance of a permanent, leak-proof joint.

When possible, the pipeline should be assembled, wholly or in sections, above ground and then lowered into a prepared trench. Bottom fill, 4 to 6 inches below pipe, should be free or rocks and other sharp objects. The same type of fill is recommended for the first 8 to 12 inches of backfill and should fully enclose the line. All lines should be laid below frost level.

Some manufacturers advise running cool water through pipe in order to shrink to normal length before and during backfilling process, especially in hot weather.

Thermal Expansion

PVC like all thermoplastic pipe materials has a relatively high rate of expansion due to temperature change as compared to ferrous materials. As with most metals, the operating temperature range will dictate the type of joint and the means of providing for expansion.

Thermal compensation of a solvent welded PVC pipeline is not required when operating extremes do not differ from installation temperatures by more than plus or minus 30 degrees for Type I and 25 degrees for Type II. Also, where infrequent cycling due to process or air temperature fluctuations occurs, threaded joints, under limitations discussed, are effective. However, solvent welded or hot air welded connections are preferable to threaded joints where any appreciable temperature changes are expected. In both cases, axial restraint guides to force the thrust along the axis of the pipe are required.

To compensate for thermal expansion and contraction, any commercial expansion joints may be used. In addition, U-bends or offsets made from hot-formed PVC pipe, or from PVC fittings and straight lengths of pipe, are suitable. If enough space is available, this latter method is the most advisable since it offers the advantages of a continuous PVC pipeline with no

Dimensions and Pressure Limits of Polyvinyl Chloride Pipe

Nominal Size, Inches	Inside Diameter, Inches	Type I - PVC Maximum Operating Pressure, Psi		Type II - PVC Maximum Operating Pressure, Psi	
		75 Deg. F	150 Deg. F	75 Deg. F	130 Deg. F
Schedule A - Lightweight					
½	0.750	165	90	145	40
¾	0.940	145	80	125	35
1	1.195	130	73	110	30
1¼	1.520	115	65	100	25
1½	1.740	115	65	100	25
2	2.175	115	65	100	25
2½	2.635	115	65	100	25
3	3.220	115	65	100	25
4	4.100	115	65	100	25
Schedule 40					
½	0.622	410	220	335	90
¾	0.824	335	180	275	70
1	1.049	310	170	255	70
1¼	1.380	255	140	210	50
1½	1.610	230	125	190	45
2	2.067	195	110	160	45
2½	2.469	200	110	165	50
3	3.068	185	100	150	40
4	4.026	155	85	130	30
6	6.065	125	65	105	30

Dimensions and Pressure Limits of Polyvinyl Chloride Pipe

Nominal Size, Inches	Inside Diameter, Inches	Type I - PVC Maximum Operating Pressure, Psi		Type II - PVC Maximum Operating Pressure, Psi	
		75 Deg. F	150 Deg. F	75 Deg. F	130 Deg. F
Schedule 80					
½	0.546	575	310	470	120
¾	0.742	470	250	385	100
1	0.957	435	235	355	95
1¼	1.278	360	195	295	80
1½	1.500	325	175	265	70
2	1.939	280	150	230	60
2½	2.323	270	140	225	60
3	2.900	260	140	215	55
4	3.826	225	120	185	50
6	5.761	195	110	160	40
Schedule 120					
½	0.500	680	360	560	145
¾	0.710	520	275	430	110
1	0.915	485	255	400	105
1¼	1.230	405	215	335	85
1½	1.450	365	190	300	80
2	1.875	320	170	265	70
2½	2.275	320	170	265	70
3	2.800	305	160	250	65
4	3.624	295	155	245	65
6	5.501	255	145	210	55

Note: Outside diameters conform to iron pipe sizes. Values of operating pressure are for plain-end pipe; values for threaded pipe would be approximately 55% in each case.

moving parts and no change in quality of corrosion resistance.

The PVC line should be installed using the same expansion offset practices as would be followed for a steel line expanding to the same extent, in inches per 100 ft.

Polyethylene

Polyethylene pipe has received, perhaps, more research and development than any other type of thermoplastic, and is currently available in three commercial grades. The points of division are based on density of the material, and the ranges are as follows:

Type I - Low Density
  Specific Gravity: 0.910 to 0.925

Type II - Intermediate
  Density
  Specific Gravity: 0.926 to 0.940

Type III - High Density
  Specific Gravity: 0.941 to 0.965

The material is manufactured from virgin polyethylene resins to which carbon black has been added for ultraviolet ray resistance. Perhaps the greatest advantage of polyethylene lies in its flexibility; low-density pipe is available in coils up to 400-ft. long through 3-inches OD (2-inches OD for high-density polyethylene). Four and six-inch sizes are available in straight lengths to 25 ft.

While high hopes are held for practical methods of joining polyethylene pipe on the outside diameter which would make it more acceptable for process piping in industry, the present joining method is by means of plastic (high-impact styrene) fittings and stainless steel clamps. Ells, tees, couplings, and reducers for joining lengths of polyethylene pipe are available, as well as threaded adapters for joining the plastic to steel pipe.

Dimensions and Pressure Limits of Type I Polyethylene Pipe

Nominal Size, Inches	Outside Diameter, Inches	Inside Diameter, Inches	Max. Operating Pressure at 75 Deg. F, Psi	Max. Operating Pressure at 120 Deg. F, Psi
75 Psi Pressure-Rated				
½	0.782	0.622	75	48
¾	1.024	0.824	75	48
1	1.300	1.050	75	48
1¼	1.710	1.380	75	48
1½	2.000	1.610	75	48
2	2.567	2.067	75	48
3	3.776	3.068	75	48
4	4.956	4.026	75	48
Schedule 40				
½	0.840	0.622	100	65
¾	1.050	0.824	83	53
1	1.315	1.049	78	50
1¼	1.660	1.380	70	45
1½	1.900	1.610	60	38
2	2.375	2.067	50	32
2½	2.875	2.469	50	32
3	3.500	3.068	50	32
4	4.500	4.026	40	25
6	6.625	6.065	35	23

Dimensions and Pressure Limits of Type I Polyethylene Pipe

Nominal Size, Inches	Outside Diameter, Inches	Inside Diameter, Inches	Max. Operating Pressure at 75 Deg. F, Psi	Max. Operating Pressure at 120 Deg. F, Psi
100 Psi Pressure-Rated				
½	0.842	0.622	100	65
¾	1.114	0.824	100	65
1	1.410	1.050	100	65
1¼	1.860	1.380	100	65
1½	2.170	1.610	100	65
2	2.777	2.067	100	65
3	4.068	3.068	100	65
4	5.386	4.026	100	65

Note: Intermediate values of operating pressure may be found with reasonable accuracy by interpolation.

For overhead runs of polyethylene pipe, practically continuous support is recommended. As with any type of plastic pipe, supports should have a reasonably broad contact surface. Wires and rods are not recommended if in direct contact with the pipe.

When laying polyethylene pipe in ditches for permanent installations, the pipe should not be tightly stretched but should be snaked at least one foot per 100 ft. to allow for thermal contraction and expansion. Ditch should then be backfilled to a minimum height of 6 inches from the top of pipe with loose material free from large rocks which might crush or cut the pipe. Normal backfilling operations may then be followed.

Dimensions and Pressure Limits of
Type III Polyethylene Pipe

Nominal Size, Inches	Outside Diameter, Inches	Inside Diameter, Inches	Max. Operating Pressure at 73 Deg. F, Psi	Max. Operating Pressure at 150 Deg. F, Psi
75 Psi Pressure-Rated				
½	0.774	0.662	100	—
¾	0.964	0.824	75	—
1	1.225	1.049	75	—
1¼	1.612	1.380	75	—
1½	1.884	1.610	75	—
2	2.375	2.067	75	—
Schedule 40				
½	0.840	0.622	150	58
¾	1.050	0.824	130	49
1	1.315	1.049	120	45
1¼	1.660	1.380	100	38
1½	1.900	1.610	90	34
2	2.375	2.067	75	29

Note: Intermediate values of operating pressure may be found with reasonable accuracy by interpolation.

Acrylonitrile-Butadiene-Styrene (ABS)

This resin-rubber blend provides the lightest weight of all semi-rigid or rigid thermoplastic pipes. ABS is supplied in 20-foot lengths in ½ to 6-inch diameters, and is available in Schedules 40, 80, and 120.

A variety of threaded and socket-type fittings, flanges, and valves are produced in ABS to give the

Dimensions and Pressure Limits of ABS Pipe

Nominal Size, Inches	Outside Diameter, Inches	Inside Diameter, Inches	Max. Operating Pressure at 70 Deg. F, Psi	Max. Operating Pressure at 170 Deg. F, Psi
Schedule 40				
½	0.840	0.622	150	75
¾	1.050	0.844	150	75
1	1.315	1.049	125	60
1¼	1.660	1.380	100	50
1½	1.900	1.610	90	45
2	2.375	2.067	75	40
Schedule 80				
½	0.840	0.546	300	150
¾	1.050	0.742	300	150
1	1.315	0.957	250	125
1¼	1.660	1.278	200	100
1½	1.900	1.500	175	90
2	2.375	1.939	150	75

pipe a wide range of application within its limits of chemical resistance.

Joining methods consist of solvent welding and threading, procedures for which would be the same as those described for polyvinyl chloride pipe. Bending of ABS as well as recommendations for buried lines, thermal expansion, and pipe supporting would also be essentially the same as those outlined for PVC.

Cellulose Acetate Butyrate

Butyrate pipe is manufactured from a product of the chemical processing of cellulose acetate and butyric acids. Although available and sometimes used in clear form for observation of flow, it normally has a non-toxic black pigment added to provide ultra-violet ray resistance for outdoor installation.

Ditching, for underground installations, does not require exacting dimensions. The semi-rigidity of this pipe permits its handling in straight lengths, yet provides sufficient flexibility so that the pipe follows contour of the ground.

When making underground installations, it is recommended that the pipe be snaked in the ditch. For each 100 feet of butyrate pipe, a minimum of $\frac{1}{8}$ inch per degree of temperature differential (that pipe will be subject to) should be allowed for thermal expansion or contraction. A concrete or dirt backfill should be poured shortly after the pipe has been placed.

Butyrate pipe is extruded in nominal diameters from $\frac{1}{2}$ through 6 inches in both Schedule 40 and Type SWP (solvent welded pipe) sizes. Solvent welded fittings in a wide variety permit easy makeup of any type of piping arrangement, including plastic-to-plastic and plastic-to-metal. The procedure would be the same as that outlined for PVC. Threading of butyrate can be done but it is not generally recommended. Bending of butyrate pipe is done in essentially the same manner as was described for PVC.

Applications of Plastic Pipe

The major application of plastic pipe in general has been in jet pump installations and other water system piping including water services, sprinkling, irrigation, and plumbing drainage. Other fields served well by thermoplastics according to their characteristics and limitations include food, beverage, and chemical pro-

Dimensions and Pressure Limits of Butyrate Pipe

Nominal Size, Inches	Outside Diameter, Inches	Inside Diameter, Inches	Max. Operating Pressure at 73 Deg. F, Psi	Max. Operating Pressure at 150 Deg. F, Psi
Type SWP (Solvent Welded Type Sizes)				
½	0.600	0.500	140	70
¾	0.855	0.750	100	50
1	1.140	1.000	100	50
1¼	1.420	1.250	100	50
1½	1.730	1.500	100	50
2	2.250	2.000	95	45
2½	2.570	2.320	80	40
3	3.250	3.000	65	30
4	4.100	3.800	60	30
6	6.220	5.760	60	30
Schedule 40				
½	0.840	0.622	220	110
¾	1.050	0.824	180	90
1	1.315	1.049	170	85
1¼	1.660	1.380	140	70
1½	1.900	1.610	130	60
2	2.375	2.067	110	55
2½	2.875	2.469	110	55
3	3.500	3.068	105	50
4	4.500	4.026	90	45
6	6.625	6.065	70	35

Note: Intermediate values of operating pressure may be found with reasonable accuracy by interpolation.

Recommended Support Spacing, in Feet, for Polyvinyl Chloride Pipe

Operating Temperature, Deg. F	Nominal Pipe Size, Inches					
	½ - ¾	1 - 1¼	1½ - 2	3	4	6
Schedule 40						
60	5.5	6.1	6.5	7.7	8.0	8.7
100	4.8	5.4	5.7	6.8	7.1	7.7
120	4.2	4.5	4.8	5.8	6.0	6.4
130	3.7	4.0	4.3	5.1	5.4	5.8
140	3.2	3.5	3.7	4.5	4.7	5.0
Schedule 80						
60	6.5	7.3	7.7	9.0	9.6	10.8
100	5.7	6.4	6.8	8.0	8.4	9.5
120	4.8	5.5	5.8	6.8	7.1	8.0
130	4.3	4.8	5.1	6.0	6.4	7.2
140	3.7	4.2	4.5	5.2	5.6	6.3
Schedule 120						
60	6.8	7.7	8.2	9.8	11.0	12.3
100	6.0	6.8	7.2	8.6	9.6	10.9
120	5.0	5.8	6.1	7.2	8.1	9.2
130	4.5	5.1	5.5	6.5	7.3	8.2
140	3.9	4.5	4.8	5.6	6.3	7.1

Note: Spacings apply to uninsulated lines carrying fluids up to 1.35 gravity. For insulated lines, reduce spans by 30%. For short spacings, economy of substituting continuous support should be considered.

Recommended Support Spacing, in Feet for Plastic Pipe Other Than PVC

Nominal Pipe Size, Inches	Operating Temperature, Deg. F		
	70	100	120
½	3.0	3.0	2.0
¾	3.5	3.0	2.5
1	3.5	3.0	2.5
1¼	4.0	3.5	3.0
1½	4.5	4.0	3.5
2	5.0	4.5	4.0
2½	6.0	5.0	4.5
3	7.0	6.0	5.0
4	8.0	7.0	6.0
6	9.0	8.0	7.0

Note: The above tables is based on Schedules SWP and 40 butyrate and Schedule 40 ABS plastic pipe, solvent welded, uninsulated, and carrying water.

For polyethylene pipe, decrease spacing by 50%,

For Schedule 80 ABS, spacing may be increased 20%.

For insulated lines, reduce spans by 30%.

For short spacings, economy of continuous support should be considered.

cess piping for industry; air conditioning recirculating lines; swimming pools; skating rinks; and natural gas and oil field piping.

Radiant heating has been and still is considered a borderline application of some of the thermoplastics. A number of systems utilizing plastic pipe have been installed for radiant heating; many have been successful, some have been troublesome. Manufacturer reluctance to positively recommend any thermoplastic for radiant heating is based not on lack of knowledge of the capability of the particular material involved, but rather on the possibility that any overheating created in the system would cause softening and possibly failure of the pipe.

Resistance to Chemical Corrosion

Corrosion of metals is usually of the galvanic or electrochemical type, characterized by a minute flow of electrical current from anodic to cathodic areas. This is accompanied by a loss of metal to the surrounding environment. Consequently, corrosion rates can easily be measured by weight loss.

With plastics, corrosion takes place in a different manner. Since plastics are non-conductors, galvanic and electrochemical effects are non-existent. Pitting and grooving do not take place, and there is no loss of material to the corrosion-causing fluid. Plastic corrosion is an "absorption" type of reaction in which the corrosive media actually penetrates or diffuses into the plastic. For this reason, corrosion is normally associated with a weight gain rather than weight loss.

Like metals, thermoplastic resistance to corrosion is influenced by concentration, temperature, and stress. The ability of the principal thermoplastics to handle various corrosive fluids is indicated in a table which follows. Comparative ratings are also given for several metallic pipe materials. Resistance ratings are divided into three categories: S, or satisfactory, indicates that chemical attack is zero or negligible; L, or limited, means the particular application should be verified with a manufacturer or tested independently; U, or unsatisfactory, indicates that the material is not recommended for the particular application. All ratings should be used for guide purposes only.

Characteristics of Pipe Plastics

Polyvinyl Chloride (PVC)	
Type I—Normal Impact	
Characteristics:	Rigid
	Excellent chemical resistance
	Normal impact strength
	Highest tensile strength of all thermoplastic pipe
	Wide operating temperature range
	Good weathering properties
	But:
	Greater weight (50% heavier than polyethylene)
Joining Methods:	Solvent welding
	Threading
	Hot welding
Applications:	Water service lines
	Process piping for chemicals
	Sour crude oil piping
	Salt water piping
Type II—High Impact	
Characteristics:	Rigid
	Improved impact strength
	Good-to-excellent chemical resistance
	Tough, even at low temperatures
	But:
	Less chemical resistance than Type I - PVC
Joining Methods:	Solvent welding
	Threading
	Hot welding
Applications:	Services requiring rough handling, shock loads

Characteristics of Pipe Plastics

Polyethylene	
Type I—Low Density	
Characteristics:	Flexible Good-to-excellent chemical resistance Lightest of all thermoplastic pipe Resists freezing Lowest in cost But: Low working pressures, operating temperatures
Joining Methods:	Insert fittings and clamps
Applications:	Water service lines Jet well systems Drainage Sprinkling and irrigation Piping salt water
Type II—Intermediate Density	
Characteristics:	Characteristics generally fall between those of low-density and high-density polyethylene
Joining Methods:	Insert fittings and clamps
Applications:	Same as low density, plus possibly process piping
Type III—High Density	
Characteristics:	Semi-flexible Good-to-excellent chemical resistance Higher tensile strength Higher impact strength
Joining Methods:	Insert fittings and clamps
Applications:	Same as low density, plus possibly process piping

Characteristics of Pipe Plastics

Acrylonitrile-Butadiene-Styrene (ABS)

Characteristics:	Rigid
	Lightest of rigid types
	Fair-to-good chemical resistance
	High heat resistance
	High impact strength
	    But:
	Lower tensile strength
	    (about midway between PVC
	    and Type I polyethylene)

Joining Methods:	Solvent welding Threading
Applications:	Water service lines Natural gas lines Oil field piping Salt water lines Process piping Sewers and drainage

Cellulose Acetate Butyrate

Characteristics:	Semi-rigid Good impact resistance Good weathering properties But: Low-to-fair chemical resistance
Joining Methods:	Solvent welding
Applications:	Water service lines Sprinkling and irrigation Sewage piping Salt water piping Natural gas lines Oil field piping

Chemical Resistance Comparison Chart—
Thermoplastics

| Key:
S—Satisfactory
L—Limited
U—Unsatisfactory | Polyvinyl Chloride | | | | Poly-ethylene | | ABS | | Butyrate | |
| | Type I | | Type II | | | | | | | |
	72°	Hot	72°	Hot	72°	Hot	72°	Hot	72°	Hot
Acetic Acid, 10%	S	S	S	L	S	S	S	S	L	—
Acetic acid, glacial	S	U	L	U	L	U	U	U	U	—
Acetone	U	U	U	U	U	U	U	U	U	U
Alcohol, methyl	S	S	S	S	S	U	S	U	U	U
Ammonium chloride	S	S	S	S	S	S	S	S	S	U
Ammonium sulfate	S	S	S	S	S	S	S	S	S	U
Aniline	U	U	U	U	U	U	U	U	U	U
Benzene	U	U	U	U	U	U	U	U	U	U
Benzoic acid	S	S	S	S	S	S	S	S	U	U
Boric acid	S	S	S	S	S	S	S	S	S	—
Butyric acid	S	U	L	U	U	U	U	U	U	—
Calcium chloride	S	S	S	S	S	S	S	S	S	—
Calcium hydroxide	S	S	S	S	S	S	S	S	L	U
Carbon tetrachloride	L	U	U	U	U	U	U	U	L	U
Chlorine, dry	S	S	S	S	L	U	S	U	U	U
Chlorine, wet	S	L	L	L	U	U	S	U	U	U
Chloroform	U	U	U	U	U	U	U	U	U	U
Chromic acid, 10%	S	S	S	S	S	S	S	U	U	U
Chromic acid, 50%	S	L	L	U	S	U	U	U	U	U
Citric acid	S	S	S	S	S	S	S	S	U	U
Copper chloride	S	S	S	S	S	S	S	S	S	—
Copper sulfate	S	S	S	S	S	S	S	S	S	U
Ethyl acetate	U	U	U	U	L	U	U	U	U	U
Ethyl chloride	U	U	U	U	U	U	U	U	U	U
Fatty acids	S	S	S	S	U	U	S	S	U	U
Ferric chloride	S	S	S	S	S	S	S	S	S	U
Ferric sulfate	S	S	S	S	S	S	S	S	S	U
Formaldehyde	S	S	S	L	S	S	S	S	U	U
Formic acid	S	U	S	U	S	L	U	U	L	U
Hydrochloric acid	S	S	S	S	S	S	S	U	U	U

Note: *Hot* refers to fluids at or near upper temperature limit
of pipe.

Chemical Resistance Comparison Chart— Metals

Key: S—Satisfactory L—Limited U—Unsatisfactory	Low Carbon Steel	Stainless Steels			Cast Iron	Aluminum	Copper	Red Brass	Lead	Monel
		302 303 304	316	410 416 430						
Acetic Acid, 10%	U	S	S	L	U	S	U	—	—	S
Acetic acid, glacial	U	S	S	U	U	S	U	U	—	—
Acetone	S	S	S	U	S	S	S	S	S	S
Alcohol, methyl	S	S	S	S	S	L	S	—	S	S
Ammonium chloride	L	L	S	L	U	U	L	—	S	S
Ammonium sulfate	S	S	S	L	S	L	L	L	—	S
Aniline	U	S	S	S	L	U	—	—	—	—
Benzene	L	S	S	—	S	S	S	S	—	S
Benzoic acid	—	S	S	S	—	S	—	—	—	—
Boric acid	U	S	S	L	L	S	S	L	S	S
Butyric acid	—	S	S	—	—	S	—	—	—	S
Calcium chloride	S	S	L	L	S	U	L	L	—	S
Calcium hydroxide	S	S	S	S	S	L	—	—	L	S
Carbon tetrachloride	L	S	S	S	L	L	S	S	L	S
Chlorine, dry	L	S	S	S	—	U	S	S	—	S
Chlorine, wet	U	U	L	U	U	U	L	U	—	L
Chloroform	—	S	S	S	—	U	—	—	—	S
Chromic acid, 10%	—	L	S	U	U	L	U	U	—	S
Chromic acid, 50%	—	U	L	—	U	—	U	U	—	—
Citric acid	U	S	S	L	L	S	S	L	—	S
Copper chloride	—	S	L	L	—	U	L	—	S	L
Copper sulfate	U	S	S	S	L	U	L	U	—	U
Ethyl acetate	L	S	S	S	S	L	—	—	—	S
Ethyl chloride	S	S	S	S	U	—	S	—	—	S
Fatty acids	—	S	S	S	—	S	—	—	—	S
Ferric chloride	U	U	U	U	U	U	U	U	—	U
Ferric sulfate	U	S	S	S	U	U	U	U	—	L
Formaldehyde	L	S	S	S	—	S	S	S	—	S
Formic acid	U	S	S	U	—	U	S	U	—	S
Hydrochloric acid	U	U	U	U	U	U	—	U	—	—

Note: Ratings for metals apply to fluid temperatures up to 140 degrees F.

Chemical Resistance Comparison Chart—
Thermoplastics

Key: S—Satisfactory L—Limited U—Unsatisfactory	Polyvinyl Chloride				Poly-ethylene		ABS		Butyrate	
	Type I		Type II							
	72°	Hot	72°	Hot	72°	Hot	72°	Hot	72°	Hot
Hydrofluoric acid	S	L	S	U	S	L	S	U	U	U
Hydrocyanic acid	S	S	S	S	S	S	S	S	U	U
Magnesium chloride	S	S	S	S	S	S	S	S	S	U
Magnesium sulfate	S	S	S	S	S	S	S	S	S	—
Nickel chloride	S	S	S	S	S	S	S	S	S	U
Nickel sulfate	S	S	S	S	S	S	S	S	S	U
Nitric acid, 20%	S	L	S	L	S	S	U	U	U	U
Nitric acid, 40%	S	S	S	L	S	L	U	U	U	U
Nitric acid, 68%	S	U	L	U	S	U	U	U	U	U
Oleic acid	S	S	S	S	L	U	S	S	L	U
Oxalic acid	S	S	S	S	S	S	S	S	U	U
Phosphoric acid, 25%	S	S	S	S	S	S	S	U	U	U
Phosphoric acid, 85%	S	S	S	S	S	L	S	U	U	U
Picric acid	U	U	U	U	S	L	S	S	U	—
Potassium chloride	S	S	S	S	S	S	S	S	S	U
Potassium sulfate	S	S	S	S	S	S	S	S	S	—
Sodium carbonate	S	S	S	S	S	S	S	S	S	—
Sodium chloride	S	S	S	S	S	S	S	S	S	—
Sodium sulfate	S	S	S	S	S	S	S	S	S	U
Stearic acid	S	S	S	S	L	L	S	S	S	—
Sulfur dioxide, dry	S	S	S	S	S	S	S	S	U	—
Sulfur dioxide, wet	S	L	U	U	S	S	S	L	U	—
Sulfuric acid, 10%	S	S	S	S	S	S	S	U	U	U
Sulfuric acid, 75%	S	S	S	S	S	L	L	U	U	U
Sulfuric acid, 90%	S	S	L	L	U	U	U	U	U	U
Trichloroethylene	U	U	U	U	U	U	U	U	U	U
Trisodium phosphate	S	S	S	S	S	S	S	S	S	—
Water, fresh	S	S	S	S	S	S	S	S	S	—
Water, salt	S	S	S	S	S	S	S	S	S	—
Zinc chloride	S	S	S	S	S	S	S	S	L	U

Note: *Hot* refers to fluids at or near upper temperature limit
of pipe.

Chemical Resistance Comparison Chart— Metals

Key: S—Satisfactory L—Limited U—Unsatisfactory	Low Carbon Steel	Stainless Steels			Cast Iron	Aluminum	Copper	Red Brass	Lead	Monel
		302 303 304	316	410 416 430						
Hydrofluoric acid	U	U	U	U	U	U	L	L	S	S
Hydrocyanic acid	L	S	S	S	S	S	L	—	—	S
Magnesium chloride	L	S	S	L	—	L	S	L	—	S
Magnesium sulfate	L	S	S	S	S	S	S	S	—	S
Nickel chloride	U	L	S	U	—	U	L	U	—	S
Nickel sulfate	U	S	S	L	—	U	U	U	—	S
Nitric acid, 20%	—	—	—	—	U	U	U	U	U	L
Nitric acid, 40%	—	—	—	—	U	U	U	U	U	U
Nitric acid, 68%	—	—	—	—	U	L	U	U	U	U
Oleic acid	L	S	S	S	—	L	S	—	—	S
Oxalic acid	U	S	S	L	U	S	S	L	—	S
Phosphoric acid, 25%	U	L	S	U	U	U	S	U	S	S
Phosphoric acid, 85%	U	U	S	U	U	U	S	U	S	S
Picric acid	—	S	S	S	—	—	U	U	—	U
Potassium chloride	—	S	S	S	—	L	S	L	—	S
Potassium sulfate	L	S	S	S	S	S	S	S	S	S
Sodium carbonate	S	S	S	S	S	U	S	L	—	S
Sodium chloride	L	S	S	L	S	L	S	—	—	S
Sodium sulfate	L	S	S	S	S	S	S	S	S	S
Stearic acid	L	S	S	S	L	S	S	L	—	S
Sulfur dioxide, dry	S	S	S	S	S	—	S	L	—	S
Sulfur dioxide, wet	—	S	S	U	—	—	S	U	—	U
Sulfuric acid, 10%	U	U	S	U	U	L	U	U	S	L
Sulfuric acid, 75%	U	U	U	U	U	U	U	U	S	S
Sulfuric acid, 90%	L	L	S	L	L	U	U	U	S	U
Trichloroethylene	L	S	S	S	L	S	S	L	—	S
Trisodium phosphate	S	S	S	S	—	U	U	U	S	S
Water, fresh	L	S	S	S	—	S	S	S	S	S
Water, salt	L	S	S	L	S	S	L	L	S	S
Zinc chloride	U	U	S	U	L	U	L	—	S	S

Note: Ratings for metals apply to fluid temperatures up to 140 degrees F.

Relative Physical Properties and Cost of Principal Thermoplastics

Physical Properties	PVC		ABS
	Type I	Type II	
Specific Gravity	1.38	1.35	1.06
Tensile Strength at 75 deg F, psi	7000	6000	4500
Modulus of Elasticity in Tension, psi $\times$ 10^5	4.15	3.50	3.00
Flexural Strength, psi	14500	11500	8000
Maximum Temperature, deg F	160	150	160
Coefficient of Expansion, inch/inch/deg F, $\times$ 10^{-5}	2.8	5.5	3.5
Thermal Conductivity, $\times$ 10^{-4} Btu/sec/sq ft/deg F/inch	3.2	3.5	2.8
Specific Heat, Btu/lb/ deg F	0.25	0.28	0.32
Relative Cost to Whole- saler of 1-inch Schedule 40 pipe, based on 1.00 for:			
Black steel pipe	1.97	1.97	1.44
Galvanized steel pipe ..	1.72	1.72	1.26

Relative Physical Properties and Cost of Principal Thermoplastics, Continued

Physical Properties	Polyethylene		Butyrate
	Type I	Type III	
Specific Gravity	0.91	0.96	1.19
Tensile Strength at 75 deg F, psi	1500	3480	4300
Modulus of Elasticity in Tension, psi $\times 10^5$	1.90	1.26	1.30
Flexural Strength, psi	1700	3700	7000
Maximum Temperature, deg F	120	150	150
Coefficient of Expansion, inch/inch/deg F, $\times$ 10^{-5}	9.4	10.0	6.5
Thermal Conductivity, $\times$ 10^{-4} Btu/sec/sq ft/deg F/inch	6.0	6.4	4.2
Specific Heat, Btu/lb/deg F	0.5	0.55	0.30
Relative Cost to Wholesaler of 1-inch Schedule 40 pipe, based on 1.00 for:			
Black steel pipe	0.79	1.00	1.69
Galvanized steel pipe ..	0.69	0.87	1.48

Spark Tests for Metals

This is a handy technique by which a welder may determine different types of metals through use of a grinding wheel. There are 5 simple rules to follow in these tests:

1. Use clean grinding wheel.
2. Hold light, even pressure against wheel.
3. Practice with known metal.

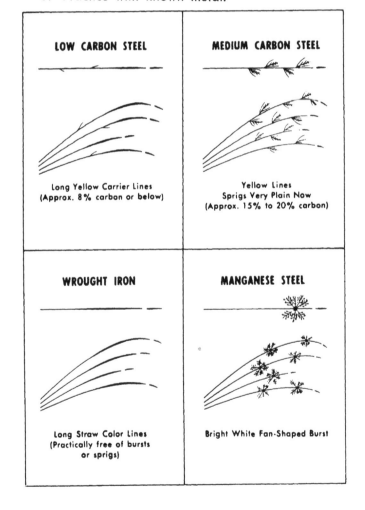

LOW CARBON STEEL

Long Yellow Carrier Lines
(Approx. 8% carbon or below)

MEDIUM CARBON STEEL

Yellow Lines
Sprigs Very Plain Now
(Approx. 15% to 20% carbon)

WROUGHT IRON

Long Straw Color Lines
(Practically free of bursts
or sprigs)

MANGANESE STEEL

Bright White Fan-Shaped Burst

Spark Tests for Metals
(Continued)

4. For best results, use at least 5000 surface feet per minute on grinding equipment.

$$\frac{\text{Inches wheel circum.} \times \text{rpm}}{12} = \text{Surface feet per minute}$$

5. Look for (A) color of stream, and (B) shape of bursts or sprigs.

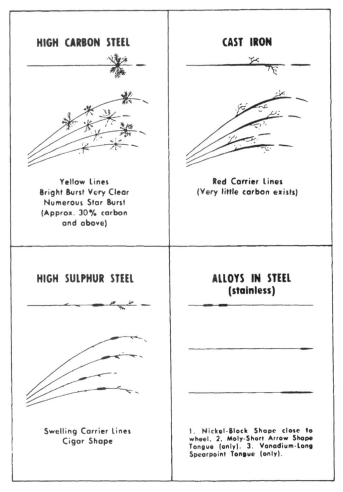

HIGH CARBON STEEL

Yellow Lines
Bright Burst Very Clear
Numerous Star Burst
(Approx. 30% carbon
and above)

CAST IRON

Red Carrier Lines
(Very little carbon exists)

HIGH SULPHUR STEEL

Swelling Carrier Lines
Cigar Shape

ALLOYS IN STEEL
(stainless)

1. Nickel-Block Shape close to
wheel. 2. Moly-Short Arrow Shape
Tongue (only). 3. Vanadium-Long
Spearpoint Tongue (only).

Gage, Thickness and Weight of Steel Sheets

Gage	Thick-ness, Inches	Steel	Galvanized Steel
		Weight, Pounds per Square Foot	
7	.1793	7.500	————
8	.1644	6.875	————
9	.1494	6.250	————
10	.1345	5.625	5.7812
11	.1196	5.000	5.1562
12	.1046	4.375	4.5312
13	.0897	3.750	3.9062
14	.0747	3.125	3.2812
15	.0673	2.812	2.9687
16	.0598	2.500	2.6562
17	.0538	2.250	2.4062
18	.0478	2.000	2.1562
19	.0418	1.750	1.9062
20	.0359	1.500	1.6562
21	.0329	1.375	1.5312
22	.0299	1.250	1.4062
23	.0269	1.125	1.2812
24	.0239	1.000	1.1562
25	.0209	.875	1.0312
26	.0179	.750	.9062
27	.0164	.6875	.8437
28	.0149	.625	.7812
29	.0135	.5625	.7187
30	.0120	.500	.6562

Sheets over $\frac{1}{4}$ inch thick are termed plates.

Weights and Melting Points of Metals

Metal	Weight per Cubic Foot, Pounds	Melting Point, Deg. F.
Aluminum	168.5	1220
Antimony	413.0	1167
Barium	235.9	1562
Bismuth	610.3	520
Boron	158.2	4172
Brass: 80C., 20Z.	536.6	1823
70C., 30Z.	526.7	1706
60C., 40Z.	521.7	1652
50C., 50Z.	511.7	1616
Bronze: 90C., 10T.	547.9	1841
Cadmium	539.6	610
Calcium	96.1	1490
Chromium	432.4	2939
Cobalt	543.5	2696
Copper	554.7	1981
Gold	1204.3	1945
Iron, cast	438.7-482.4	2000
Iron, wrought	486.7-493.0	2750
Lead	707.7	621
Magnesium	108.6	1204
Manganese	455.5	2300
Mercury	845.3	−38
Molybdenum	636.5	4748
Nickel	549.1	2651
Platinum	1333.5	3224
Potassium	54.3	144
Silver	650.2-657.1	1761
Sodium	60.6	207
Steel, Carbon	489.0-490.8	2500
Tantalum	1035.8	5162
Tin	454.9	449
Titanium	280.1	3272
Tungsten	1161-1192	6098
Uranium	1166.9	3362
Vanadium	394.4	3110
Zinc	439.3-446.8	788

Temperature of Saturated Steam

Absolute Pressure, Lb. per Sq. In.	Vacuum in Inches of Mercury	Temperature, Deg. F.
0.0886	29.74	32
0.1217	29.67	40
0.2562	29.40	60
0.505	29.89	80
0.946	28.00	100
1	27.88	101.83
2	25.85	126.15
3	23.81	141.52
4	21.78	153.01
5	19.74	162.28
10	9.56	193.22

Absolute Pressure, Lb. per Sq. In.	Gage Pressure, Lb. per Sq. In.	Temperature, Deg. F.
14.7	0	212.0
15	0.3	213.0
20	5.3	228.0
25	10.3	240.1
30	15.3	250.3
40	25.3	267.3
50	35.3	281.0
100	85.3	327.8
150	135.3	358.5
200	185.3	381.9
250	235.3	401.1
300	285.3	417.5
400	385.3	444.8
500	485.3	467.3
600	583.3	486.6

Circumferences and Areas of Circles

Diam-eter	Circum-ference	Area	Diam-eter	Circum-ference	Area
⅛	0.3927	0.0123	⅝	11.3883	10.321
¼	0.7854	0.0491	¾	11.7810	11.045
⅜	1.1781	0.1105	⅞	12.1737	11.793
½	1.5708	0.1964	4	12.5664	12.566
⅝	1.9635	0.3068	⅛	12.9591	13.364
¾	2.3562	0.4418	¼	13.3518	14.186
⅞	2.7489	0.6013	⅜	13.7445	15.033
1	3.1416	0.7854	½	14.1372	15.904
⅛	3.5343	0.9940	⅝	14.5299	16.800
¼	3.9270	1.2272	¾	14.9226	17.721
⅜	4.3197	1.4849	⅞	15.3153	18.665
½	4.7124	1.7671	5	15.7080	19.635
⅝	5.1051	2.0739	⅛	16.1007	20.629
¾	5.4978	2.4053	¼	16.4934	21.648
⅞	5.8905	2.7612	⅜	16.8861	22.691
2	6.2832	3.1416	½	17.2788	23.758
⅛	6.6759	3.5466	⅝	17.6715	24.850
¼	7.0686	3.9761	¾	18.0642	25.967
⅜	7.4613	4.4301	⅞	18.4569	27.109
½	7.8540	4.9087	6	18.8496	28.274
⅝	8.2467	5.4119	⅛	19.2423	29.465
¾	8.6394	5.9396	¼	19.6350	30.680
⅞	9.0321	6.4918	⅜	20.0277	31.919
3	9.4248	7.0686	½	20.4204	33.183
⅛	9.8175	7.6699	⅝	20.8131	34.472
¼	10.2102	8.2958	¾	21.2058	35.785
⅜	10.6029	8.9462	⅞	21.5984	37.122
½	10.9956	9.6211	7	21.9911	38.485

Circumferences and Areas of Circles

Diam-eter	Circum-ference	Area	Diam-eter	Circum-ference	Area
7	21.9911	38.485	11	34.5575	95.03
$\frac{1}{8}$	22.3838	39.871	$\frac{1}{8}$	34.9502	97.20
$\frac{1}{4}$	22.7765	41.282	$\frac{1}{4}$	35.3429	99.40
$\frac{3}{8}$	23.1692	42.718	$\frac{3}{8}$	35.7356	101.62
$\frac{1}{2}$	23.5619	44.179	$\frac{1}{2}$	36.1283	103.87
$\frac{5}{8}$	23.9546	45.664	$\frac{5}{8}$	36.5210	106.14
$\frac{3}{4}$	24.3473	47.173	$\frac{3}{4}$	36.9137	108.43
$\frac{7}{8}$	24.7400	48.707	$\frac{7}{8}$	37.3064	110.75
8	25.1327	50.265	12	37.6991	113.10
$\frac{1}{8}$	25.5254	51.849	$\frac{1}{8}$	38.0918	115.47
$\frac{1}{4}$	25.9181	53.456	$\frac{1}{4}$	38.4845	117.86
$\frac{3}{8}$	26.3108	55.088	$\frac{3}{8}$	38.8772	120.28
$\frac{1}{2}$	26.7035	56.745	$\frac{1}{2}$	39.2699	122.72
$\frac{5}{8}$	27.0962	58.426	$\frac{5}{8}$	39.6626	125.19
$\frac{3}{4}$	27.4889	60.132	$\frac{3}{4}$	40.0553	127.68
$\frac{7}{8}$	27.8816	61.862	$\frac{7}{8}$	40.4480	130.19
9	28.2743	63.617	13	40.8407	132.73
$\frac{1}{8}$	28.6670	65.397	$\frac{1}{8}$	41.2334	135.30
$\frac{1}{4}$	29.0597	67.201	$\frac{1}{4}$	41.6261	137.89
$\frac{3}{8}$	29.4524	69.029	$\frac{3}{8}$	42.0188	140.50
$\frac{1}{2}$	29.8451	70.882	$\frac{1}{2}$	42.4115	143.14
$\frac{5}{8}$	30.2378	72.760	$\frac{5}{8}$	42.8042	145.80
$\frac{3}{4}$	30.6305	74.662	$\frac{3}{4}$	43.1969	148.49
$\frac{7}{8}$	31.0232	76.589	$\frac{7}{8}$	43.5896	151.20
10	31.4159	78.540	14	43.9823	153.94
$\frac{1}{8}$	31.8086	80.516	$\frac{1}{8}$	44.3750	156.70
$\frac{1}{4}$	32.2013	82.516	$\frac{1}{4}$	44.7677	159.48
$\frac{3}{8}$	32.5940	84.541	$\frac{3}{8}$	45.1604	162.30
$\frac{1}{2}$	32.9867	86.590	$\frac{1}{2}$	45.5531	165.13
$\frac{5}{8}$	33.3794	88.664	$\frac{5}{8}$	45.9458	167.99
$\frac{3}{4}$	33.7721	90.763	$\frac{3}{4}$	46.3385	170.87
$\frac{7}{8}$	34.1648	92.886	$\frac{7}{8}$	46.7312	173.78

Circumferences and Areas of Circles

Diam-eter	Circum-ference	Area	Diam-eter	Circum-ference	Area
15	47.1239	176.71	19	59.6903	283.53
$\frac{1}{8}$	47.5166	179.67	$\frac{1}{8}$	60.0830	287.27
$\frac{1}{4}$	47.9093	182.65	$\frac{1}{4}$	60.4757	291.04
$\frac{3}{8}$	48.3020	185.66	$\frac{3}{8}$	60.8684	294.83
$\frac{1}{2}$	48.6947	188.69	$\frac{1}{2}$	61.2611	298.65
$\frac{5}{8}$	49.0874	191.75	$\frac{5}{8}$	61.6538	302.49
$\frac{3}{4}$	49.4801	194.83	$\frac{3}{4}$	62.0465	306.35
$\frac{7}{8}$	49.8728	197.93	$\frac{7}{8}$	62.4392	310.24
16	50.2655	201.06	20	62.8319	314.16
$\frac{1}{8}$	50.6582	204.22	$\frac{1}{8}$	63.2246	318.10
$\frac{1}{4}$	51.0509	207.39	$\frac{1}{4}$	63.6173	322.06
$\frac{3}{8}$	51.4436	210.60	$\frac{3}{8}$	64.0100	326.05
$\frac{1}{2}$	51.8363	213.82	$\frac{1}{2}$	64.4026	330.06
$\frac{5}{8}$	52.2290	217.08	$\frac{5}{8}$	64.7953	334.10
$\frac{3}{4}$	52.6217	220.35	$\frac{3}{4}$	65.1880	338.16
$\frac{7}{8}$	53.0144	223.65	$\frac{7}{8}$	65.5807	342.25
17	53.4071	226.98	21	65.9734	346.36
$\frac{1}{8}$	53.7998	230.33	$\frac{1}{8}$	66.3661	350.50
$\frac{1}{4}$	54.1925	233.71	$\frac{1}{4}$	66.7588	354.66
$\frac{3}{8}$	54.5852	237.10	$\frac{3}{8}$	67.1515	358.84
$\frac{1}{2}$	54.9779	240.53	$\frac{1}{2}$	67.5442	363.05
$\frac{5}{8}$	55.3706	243.98	$\frac{5}{8}$	67.9369	367.28
$\frac{3}{4}$	55.7633	247.45	$\frac{3}{4}$	68.3296	371.54
$\frac{7}{8}$	56.1560	250.95	$\frac{7}{8}$	68.7223	375.83
18	56.5487	254.47	22	69.1150	380.13
$\frac{1}{8}$	56.9414	258.02	$\frac{1}{8}$	69.5077	384.46
$\frac{1}{4}$	57.3341	261.59	$\frac{1}{4}$	69.9004	388.82
$\frac{3}{8}$	57.7268	265.18	$\frac{3}{8}$	70.2931	393.20
$\frac{1}{2}$	58.1195	268.80	$\frac{1}{2}$	70.6858	397.61
$\frac{5}{8}$	58.5122	272.45	$\frac{5}{8}$	71.0785	402.04
$\frac{3}{4}$	58.9049	276.12	$\frac{3}{4}$	71.4712	406.49
$\frac{7}{8}$	59.2976	279.81	$\frac{7}{8}$	71.8639	410.97

Circumferences and Areas of Circles

Diameter	Circumference	Area	Diameter	Circumference	Area
23	72.2566	415.48	27	84.8230	572.56
1/8	72.6493	420.00	1/8	85.2157	577.87
1/4	73.0420	424.56	1/4	85.6084	583.21
3/8	73.4347	429.13	3/8	86.0011	588.57
1/2	73.8274	433.74	1/2	86.3938	593.96
5/8	74.2201	438.36	5/8	86.7865	599.37
3/4	74.6128	443.01	3/4	87.1792	604.81
7/8	75.0055	447.69	7/8	87.5719	610.27
24	75.3982	452.39	28	87.9646	615.75
1/8	75.7909	457.11	1/8	88.3573	621.26
1/4	76.1836	461.86	1/4	88.7500	626.80
3/8	76.5763	466.64	3/8	89.1427	632.36
1/2	76.9690	471.44	1/2	89.5354	637.94
5/8	77.3617	476.26	5/8	89.9281	643.55
3/4	77.7544	481.11	3/4	90.3208	649.18
7/8	78.1471	485.98	7/8	90.7135	654.84
25	78.5398	490.87	29	91.1062	660.52
1/8	78.9325	495.79	1/8	91.4989	666.23
1/4	79.3252	500.74	1/4	91.8916	671.96
3/8	79.7179	505.71	3/8	92.2843	677.71
1/2	80.1106	510.71	1/2	92.6770	683.49
5/8	80.5033	515.72	5/8	93.0697	689.30
3/4	80.8960	520.77	3/4	93.4624	695.13
7/8	81.2887	525.84	7/8	93.8551	700.98
26	81.6814	530.93	30	94.2478	706.86
1/8	82.0741	536.05	1/8	94.6405	712.76
1/4	82.4668	541.19	1/4	95.0332	718.69
3/8	82.8595	546.35	3/8	95.4259	724.64
1/2	83.2522	551.55	1/2	95.8186	730.62
5/8	83.6449	556.76	5/8	96.2113	736.62
3/4	84.0376	562.00	3/4	96.6040	742.64
7/8	84.4303	567.27	7/8	96.9967	748.69

Circumferences and Areas of Circles

Diam-eter	Circum-ference	Area	Diam-eter	Circum-ference	Area
31	97.3894	754.77	35	109.956	962.1
$\frac{1}{8}$	97.7821	760.87	$\frac{1}{8}$	110.348	969.0
$\frac{1}{4}$	98.1748	766.99	$\frac{1}{4}$	110.741	975.9
$\frac{3}{8}$	98.5675	773.14	$\frac{3}{8}$	111.134	982.8
$\frac{1}{2}$	98.9602	779.31	$\frac{1}{2}$	111.527	989.8
$\frac{5}{8}$	99.3529	785.51	$\frac{5}{8}$	111.919	996.8
$\frac{3}{4}$	99.7456	791.73	$\frac{3}{4}$	112.312	1003.8
$\frac{7}{8}$	100.138	797.98	$\frac{7}{8}$	112.705	1010.8
32	100.531	804.25	36	113.097	1017.9
$\frac{1}{8}$	100.924	810.54	$\frac{1}{8}$	113.490	1025.0
$\frac{1}{4}$	101.316	816.86	$\frac{1}{4}$	113.883	1032.1
$\frac{3}{8}$	101.709	823.21	$\frac{3}{8}$	114.275	1039.2
$\frac{1}{2}$	102.102	829.58	$\frac{1}{2}$	114.668	1046.3
$\frac{5}{8}$	102.494	835.97	$\frac{5}{8}$	115.061	1053.5
$\frac{3}{4}$	102.887	842.39	$\frac{3}{4}$	115.454	1060.7
$\frac{7}{8}$	103.280	848.33	$\frac{7}{8}$	115.846	1068.0
33	103.673	855.30	37	116.239	1075.2
$\frac{1}{8}$	104.065	861.79	$\frac{1}{8}$	116.632	1082.5
$\frac{1}{4}$	104.458	868.31	$\frac{1}{4}$	117.024	1089.8
$\frac{3}{8}$	104.851	874.85	$\frac{3}{8}$	117.417	1097.1
$\frac{1}{2}$	105.243	881.41	$\frac{1}{2}$	117.810	1104.5
$\frac{5}{8}$	105.636	888.00	$\frac{5}{8}$	118.202	1111.8
$\frac{3}{4}$	106.029	894.62	$\frac{3}{4}$	118.596	1119.2
$\frac{7}{8}$	106.421	901.26	$\frac{7}{8}$	118.988	1126.7
34	106.814	907.92	38	119.381	1134.1
$\frac{1}{8}$	107.207	914.61	$\frac{1}{8}$	119.773	1141.6
$\frac{1}{4}$	107.600	921.32	$\frac{1}{4}$	120.166	1149.1
$\frac{3}{8}$	107.992	928.06	$\frac{3}{8}$	120.559	1156.6
$\frac{1}{2}$	108.385	934.82	$\frac{1}{2}$	120.951	1164.2
$\frac{5}{8}$	108.778	941.61	$\frac{5}{8}$	121.344	1171.7
$\frac{3}{4}$	109.170	948.42	$\frac{3}{4}$	121.737	1179.3
$\frac{7}{8}$	109.563	955.25	$\frac{7}{8}$	122.129	1186.9

Circumferences and Areas of Circles

Diam-eter	Circum-ference	Area	Diam-eter	Circum-ference	Area
39	122.522	1194.6	43	135.088	1452.2
$\frac{1}{8}$	122.915	1202.3	$\frac{1}{8}$	135.481	1460.7
$\frac{1}{4}$	123.308	1210.0	$\frac{1}{4}$	135.874	1469.1
$\frac{3}{8}$	123.700	1217.7	$\frac{3}{8}$	136.267	1477.6
$\frac{1}{2}$	124.093	1225.4	$\frac{1}{2}$	136.659	1486.2
$\frac{5}{8}$	124.486	1233.2	$\frac{5}{8}$	137.052	1494.7
$\frac{3}{4}$	124.878	1241.0	$\frac{3}{4}$	137.445	1503.3
$\frac{7}{8}$	125.271	1248.8	$\frac{7}{8}$	137.837	1511.9
40	125.664	1256.6	44	138.230	1520.5
$\frac{1}{8}$	126.056	1264.5	$\frac{1}{8}$	138.623	1529.2
$\frac{1}{4}$	126.449	1272.4	$\frac{1}{4}$	139.015	1537.9
$\frac{3}{8}$	126.842	1280.3	$\frac{3}{8}$	139.408	1546.6
$\frac{1}{2}$	127.235	1288.2	$\frac{1}{2}$	139.801	1555.3
$\frac{5}{8}$	127.627	1296.2	$\frac{5}{8}$	140.194	1564.0
$\frac{3}{4}$	128.020	1304.2	$\frac{3}{4}$	140.586	1572.8
$\frac{7}{8}$	128.413	1312.2	$\frac{7}{8}$	140.979	1581.6
41	128.805	1320.3	45	141.372	1590.4
$\frac{1}{8}$	129.198	1328.3	$\frac{1}{8}$	141.764	1599.3
$\frac{1}{4}$	129.591	1336.4	$\frac{1}{4}$	142.157	1608.2
$\frac{3}{8}$	129.983	1344.5	$\frac{3}{8}$	142.550	1617.0
$\frac{1}{2}$	130.376	1352.7	$\frac{1}{2}$	142.942	1626.0
$\frac{5}{8}$	130.769	1360.8	$\frac{5}{8}$	143.335	1634.9
$\frac{3}{4}$	131.161	1369.0	$\frac{3}{4}$	143.728	1643.9
$\frac{7}{8}$	131.554	1377.2	$\frac{7}{8}$	144.121	1652.9
42	131.947	1385.4	46	144.513	1661.9
$\frac{1}{8}$	132.340	1393.7	$\frac{1}{8}$	144.906	1670.9
$\frac{1}{4}$	132.732	1402.0	$\frac{1}{4}$	145.299	1680.0
$\frac{3}{8}$	133.125	1410.3	$\frac{3}{8}$	145.691	1689.1
$\frac{1}{2}$	133.518	1418.6	$\frac{1}{2}$	146.084	1698.2
$\frac{5}{8}$	133.910	1427.0	$\frac{5}{8}$	146.477	1707.4
$\frac{3}{4}$	134.303	1435.4	$\frac{3}{4}$	146.869	1716.5
$\frac{7}{8}$	134.696	1443.8	$\frac{7}{8}$	147.262	1725.7

Squares and Square Roots of Numbers from 1 to 1000

Second column in this table gives the square of the number in the first column, while third column gives square root of the number in the first column.

No.	Square	Square Root	No.	Square	Square Root
1	1	1.0000	33	1089	5.7446
2	4	1.4142	34	1156	5.8310
3	9	1.7321	35	1225	5.9161
4	16	2.0000	36	1296	6.0000
5	25	2.2361	37	1369	6.0828
6	36	2.4495	38	1444	6.1644
7	49	2.6458	39	1521	6.2450
8	64	2.8284	40	1600	6.3246
9	81	3.0000	41	1681	6.4031
10	100	3.1623	42	1764	6.4807
11	121	3.3166	43	1849	6.5574
12	144	3.4641	44	1936	6.6332
13	169	3.6056	45	2025	6.7082
14	196	3.7417	46	2116	6.7823
15	225	3.8730	47	2209	6.8557
16	256	4.0000	48	2304	6.9282
17	289	4.1231	49	2401	7.0000
18	324	4.2426	50	2500	7.0711
19	361	4.3589	51	2601	7.1414
20	400	4.4721	52	2704	7.2111
21	441	4.5826	53	2809	7.8201
22	484	4.6904	54	2916	7.3485
23	529	4.7958	55	3025	7.4162
24	576	4.8990	56	3136	7.4833
25	625	5.0000	57	3249	7.5498
26	676	5.0990	58	3364	7.6158
27	729	5.1962	59	3481	7.6811
28	784	5.2915	60	3600	7.7460
29	841	5.3852	61	3721	7.8102
30	900	5.4772	62	3844	7.8740
31	961	5.5678	63	3969	7.9373
32	1024	5.6569	64	4096	8.0000

Squares and Square Roots of Numbers

No.	Square	Square Root	No.	Square	Square Root
65	4225	8.0623	101	10201	10.0499
66	4356	8.1240	102	10404	10.0995
67	4489	8.1854	103	10609	10.1489
68	4624	8.2462	104	10816	10.1980
69	4761	8.3066	105	11025	10.2470
70	4900	8.3666	106	11236	10.2956
71	5041	8.4261	107	11449	10.3441
72	5184	8.4853	108	11664	10.3923
73	5329	8.5440	109	11881	10.4403
74	5476	8.6023	110	12100	10.4881
75	5625	8.6603	111	12321	10.5357
76	5776	8.7178	112	12544	10.5830
77	5929	8.7750	113	12769	10.6301
78	6084	8.8318	114	12996	10.6771
79	6241	8.8882	115	13225	10.7238
80	6400	8.9443	116	13456	10.7703
81	6561	9.0000	117	13689	10.8167
82	6724	9.0554	118	13924	10.8628
83	6889	9.1104	119	14161	10.9087
84	7056	9.1652	120	14400	10.9545
85	7225	9.2195	121	14641	11.0000
86	7396	9.2736	122	14884	11.0454
87	7569	9.3276	123	15129	11.0905
88	7744	9.3808	124	15376	11.1355
89	7921	9.4340	125	15625	11.1803
90	8100	9.4868	126	15876	11.2250
91	8281	9.5394	127	16129	11.2694
92	8464	9.5917	128	16384	11.3137
93	8649	9.6437	129	16641	11.3578
94	8836	9.6954	130	16900	11.4018
95	9025	9.7468	131	17161	11.4455
96	9216	9.7980	132	17424	11.4891
97	9409	9.8489	133	17689	11.5326
98	9604	9.8995	134	17956	11.5758
99	9801	9.9499	135	18225	11.6190
100	10000	10.0000	136	18496	11.6619

Squares and Square Roots of Numbers

No.	Square	Square Root	No.	Square	Square Root
137	18769	11.7047	173	29929	13.1529
138	19044	11.7473	174	30276	13.1909
139	19321	11.7898	175	30625	13.2288
140	19600	11.8322	176	30976	13.2665
141	19881	11.8743	177	31329	13.3041
142	20164	11.9164	178	31684	13.3417
143	20449	11.9583	179	32041	13.3791
144	20736	12.0000	180	32400	13.4164
145	21025	12.0416	181	32761	13.4536
146	21316	12.0830	182	33124	13.4907
147	21609	12.1244	183	33489	13.5277
148	21904	12.1655	184	33856	13.5647
149	22201	12.2066	185	34225	13.6015
150	22500	12.2474	186	34596	13.6382
151	22801	12.2882	187	34969	13.6748
152	23104	12.3288	188	35344	13.7113
153	23409	12.3693	189	35721	13.7477
154	23716	12.4097	190	36100	13.7840
155	24025	12.4499	191	36481	13.8203
156	24336	12.4900	192	36864	13.8564
157	24649	12.5300	193	37249	13.8924
158	24964	12.5698	194	37636	13.9284
159	25281	12.6095	195	38025	13.9642
160	25600	12.6491	196	38416	14.0000
161	25921	12.6886	197	38809	14.0357
162	26244	12.7279	198	39204	14.0712
163	26569	12.7671	199	39601	14.1067
164	26896	12.8062	200	40000	14.1421
165	27225	12.8452	201	40401	14.1774
166	27556	12.8841	202	40804	14.2127
167	27889	12.9228	203	41209	14.2478
168	28224	12.9615	204	41616	14.2829
169	28561	13.0000	205	42025	14.3178
170	28900	13.0384	206	42436	14.3527
171	29241	13.0767	207	42849	14.3875
172	29584	13.1149	208	43264	14.4222

Squares and Square Roots of Numbers

No.	Square	Square Root	No.	Square	Square Root
209	43681	14.4568	245	60025	15.6525
210	44100	14.4914	246	60516	15.6844
211	44521	14.5258	247	61009	15.7162
212	44944	14.5602	248	61504	15.7480
213	45369	14.5945	249	62001	15.7797
214	45796	14.6287	250	62500	15.8114
215	46225	14.6629	251	63001	15.8430
216	36656	14.6969	252	63504	15.8745
217	47089	14.7309	253	64009	15.9060
218	47524	14.7648	254	64516	15.9374
219	47961	14.7986	255	65025	15.9687
220	48400	14.8324	256	65536	16.0000
221	48841	14.8661	257	66049	16.0312
222	49284	14.8997	258	66564	16.0624
223	49729	14.9332	259	67081	16.0935
224	50176	14.9666	260	67600	16.1245
225	50625	15.0000	261	68121	16.1555
226	51076	15.0333	262	68644	16.1864
227	51529	15.0665	263	69169	16.2173
228	51984	15.0997	264	69696	16.2481
229	52441	15.1327	265	70225	16.2788
230	52900	15.1658	266	70756	16.3095
231	53361	15.1987	267	71289	16.3401
232	53824	15.2315	268	71824	16.3707
233	54289	15.2643	269	72361	16.4012
234	54756	15.2971	270	72900	16.4317
235	55225	15.3297	271	73441	16.4621
236	55696	15.3623	272	73984	16.4924
237	56169	15.3948	273	74529	16.5227
238	56644	15.4272	274	75076	16.5529
239	57121	15.4596	275	75625	16.5831
240	57600	15.4919	276	76176	16.6132
241	58081	15.5242	277	76729	16.6433
242	58564	15.5563	278	77284	16.6733
243	59049	15.5885	279	77841	16.7033
244	59536	15.6205	280	78400	16.7332

Squares and Square Roots of Numbers

No.	Square	Square Root	No.	Square	Square Root
281	78961	16.7631	317	100489	17.8045
282	79524	16.7929	318	101124	17.8326
283	80089	16.8226	319	101761	17.8606
284	80656	16.8523	320	102400	17.8885
285	81225	16.8819	321	103041	17.9165
286	81796	16.9115	322	103684	17.9444
287	82369	16.9411	323	104329	17.9722
288	82944	16.9706	324	104976	18.0000
289	83521	17.0000	325	105625	18.0278
290	84100	17.0294	326	106276	18.0555
291	84681	17.0587	327	106929	18.0831
292	85264	17.0880	328	107584	18.1108
293	85849	17.1172	329	108241	18.1384
294	86436	17.1464	330	108900	18.1659
295	87025	17.1756	331	109561	18.1934
296	87616	17.2047	332	110224	18.2209
297	88209	17.2337	333	110889	18.2483
298	88804	17.2627	334	111556	18.2757
299	89401	17.2916	335	112225	18.3030
300	90000	17.3205	336	112896	18.3303
301	90601	17.3494	337	113569	18.3576
302	91204	17.3781	338	114244	18.3848
303	91809	17.4069	339	114921	18.4120
304	92416	17.4356	340	115600	18.4391
305	93025	17.4642	341	116281	18.4662
306	93636	17.4029	342	116964	18.4932
307	94249	17.5214	343	117649	18.5203
308	94864	17.5499	344	118336	18.5472
309	95481	17.5784	345	119025	18.5742
310	96100	17.6068	346	119716	18.6011
311	96721	17.6352	347	120409	18.6279
312	97344	17.6635	348	121104	18.6548
313	97969	17.6918	349	121801	18.6815
314	98596	17.7200	350	122500	18.7083
315	99225	17.7482	351	123201	18.7350
316	99856	17.7764	352	123904	18.7617

Squares and Square Roots of Numbers

No.	Square	Square Root	No.	Square	Square Root
353	124608	18.7883	389	151321	19.7231
354	125316	18.8149	390	152100	19.7484
355	126025	18.8414	391	152881	19.7737
356	126736	18.8680	392	153664	19.7990
357	127449	18.8944	393	154449	19.8242
358	128164	18.9209	394	155236	19.8494
359	128881	18.9473	395	156025	19.8746
360	129600	18.9737	396	156816	19.8997
361	130321	19.0000	397	157609	19.9249
362	131044	19.0263	398	158404	19.9499
363	131769	19.0526	399	159201	19.9750
364	132496	19.0788	400	160000	20.0000
365	133225	19.1050	401	160801	20.0250
366	133956	19.1311	402	161604	20.0499
367	134689	19.1572	403	162409	20.0749
368	135424	19.1833	404	163216	20.0998
369	136161	19.2094	405	164025	20.1246
370	136900	19.2354	406	164836	20.1494
371	137641	19.2614	407	165649	20.1742
372	138384	19.2873	408	166464	20.1990
373	139129	19.3132	409	167281	20.2237
374	139876	19.3391	410	168100	20.2485
375	140625	19.3649	411	168921	20.2731
376	141376	19.3907	412	169744	20.2978
377	142129	19.4165	413	170569	20.3224
378	142884	19.4422	414	171396	20.3470
379	143641	19.4679	415	172225	20.3715
380	144400	19.4936	416	173056	20.3961
381	145161	19.5192	417	173889	20.4206
382	145924	19.5448	418	174724	20.4450
383	146689	19.5704	419	175561	20.4695
384	147456	19.5959	420	176400	20.4939
385	148225	19.6214	421	177241	20.5183
386	148996	19.6469	422	178084	20.5426
387	149769	19.6723	423	178929	20.5670
388	150544	19.6977	424	179776	20.5913

Squares and Square Roots of Numbers

No.	Square	Square Root	No.	Square	Square Root
425	180625	20.6155	461	212521	21.4709
426	181476	20.6398	462	213444	21.4942
427	182329	20.6640	463	214369	21.5174
428	183184	20.6882	464	215296	21.5407
429	184041	20.7123	465	216225	21.5639
430	184900	20.7364	466	217156	21.5870
431	185761	20.7605	467	218089	21.6102
432	186624	20.7846	468	219024	21.6333
433	187489	20.8087	469	219961	21.6564
434	188356	20.8327	470	220900	21.6795
435	189225	20.8567	471	221841	21.7025
436	190096	20.8806	472	222784	21.7256
437	190969	20.9045	473	223729	21.7486
438	191844	20.9284	474	224676	21.7715
439	192721	20.9523	475	225625	21.7945
440	193600	20.9762	476	226576	21.8174
441	194481	21.0000	477	227529	21.8403
442	195364	21.0238	478	228484	21.8632
443	196249	21.0476	479	229441	21.8861
444	197136	21.0713	480	230400	21.9089
445	198025	21.0950	481	231361	21.9317
446	198916	21.1187	482	232324	21.9545
447	199809	21.1424	483	233289	21.9773
448	200704	21.1660	484	234256	22.0000
449	201601	21.1896	485	235225	22.0227
450	202500	21.2132	486	236196	22.0454
451	203401	21.2368	487	237169	22.0681
452	204304	21.2603	488	238144	22.0907
453	205209	21.2838	489	239121	22.1133
454	206116	21.3073	490	240100	22.1359
455	207025	21.3307	491	241081	22.1585
456	207936	21.3542	492	242064	22.1811
457	208849	21.3776	493	243049	22.2036
458	209764	21.4009	494	244036	22.2261
459	210681	21.4243	495	245025	22.2486
460	211600	21.4476	496	246016	22.2711

Squares and Square Roots of Numbers

No.	Square	Square Root	No.	Square	Square Root
497	247009	22.2935	533	284089	23.0868
498	248004	22.3159	534	285156	23.1084
499	249001	22.3383	535	286225	23.1301
500	250000	22.3607	536	287296	23.1517
501	251001	22.3830	537	288369	23.1733
502	252004	22.4054	538	289444	23.1948
503	253009	22.4277	539	290521	23.2164
504	254016	22.4499	540	291600	23.2379
505	255025	22.4722	541	292681	23.2594
506	256036	22.4944	542	293764	23.2809
507	257049	22.5167	543	294849	23.3024
508	258064	22.5389	544	295936	23.3238
509	259081	22.5610	545	297025	23.3452
510	260100	22.5832	546	298116	23.3666
511	261121	22.6053	547	299209	23.3880
512	262144	22.6274	548	300304	23.4094
513	263169	22.6495	549	301401	23.4307
514	264196	22.6716	550	302500	23.4521
515	265225	22.6936	551	303601	23.4734
516	266256	22.7156	552	304704	23.4947
517	267289	22.7376	553	305809	23.5160
518	268324	22.7596	554	306916	23.5372
519	269361	22.7816	555	309025	23.5584
520	270400	22.8035	556	309136	23.5797
521	271441	22.8254	557	310249	23.6008
522	272484	22.8473	558	311364	23.6220
523	273529	22.8692	559	312481	23.6432
524	274576	22.8910	560	313600	23.6643
525	275625	22.9129	561	314721	23.6854
526	276676	22.9347	562	315844	23.7065
527	277729	22.9565	563	316969	23.7276
528	278784	22.9783	564	318096	23.7487
529	279841	23.0000	565	319225	23.7697
530	280900	23.0217	566	320356	23.7908
531	281961	23.0434	567	321489	23.8118
532	283024	23.0651	568	322624	23.8328

Squares and Square Roots of Numbers

No.	Square	Square Root	No.	Square	Square Root
569	323761	23.8537	605	366025	24.5967
570	324900	23.8747	606	367236	24.6171
571	326041	23.8956	607	368449	24.6374
572	327184	23.9165	608	369664	24.6577
573	328329	23.9374	609	370881	24.6779
574	329476	23.9583	610	372100	24.6982
575	330625	23.9792	611	373321	24.7184
576	331776	24.0000	612	374544	24.7386
577	332929	24.0208	613	375769	24.7588
578	334084	24.0416	614	376996	24.7790
579	335241	24.0624	615	378225	24.7992
580	336400	24.0832	616	379456	24.8193
581	337561	24.1039	617	380689	24.8395
582	338724	24.1247	618	381924	24.8596
583	339889	24.1454	619	383161	24.8797
584	341056	24.1661	620	384400	24.8998
585	342225	24.1868	621	385641	24.9199
586	343396	24.2074	622	386884	24.9399
587	344569	24.2281	623	388129	24.9600
588	345744	24.2487	624	389376	24.9800
589	346921	24.2693	625	390625	25.0000
590	348100	24.2899	626	391876	25.0200
591	349281	24.3105	627	393129	25.0400
592	350464	24.3311	628	394384	25.0599
593	351649	24.3516	629	395641	25.0799
594	352836	24.3721	630	396900	25.0998
595	354025	24.3926	631	398161	25.1197
596	355216	24.4131	632	399424	25.1396
597	356409	24.4336	633	400689	25.1595
598	357604	24.4540	634	401956	25.1794
599	358801	24.4715	635	403225	25.1992
600	360000	24.4949	636	404496	25.2190
601	361201	24.5153	637	405769	25.2389
602	362404	24.5357	638	407044	25.2587
603	363609	24.5561	639	408321	25.2784
604	364816	24.5764	640	409600	25.2982

Squares and Square Roots of Numbers

No.	Square	Square Root	No.	Square	Square Root
641	410881	25.3180	677	458329	26.0192
642	412164	25.3377	678	459684	26.0384
643	413449	25.3574	679	461041	26.0576
644	414736	25.3772	680	462400	26.0768
645	416025	25.3969	681	463761	26.0960
646	417316	25.4165	682	465124	26.1151
647	418609	25.4362	683	466489	26.1343
648	419904	25.4558	684	467856	26.1534
649	421201	25.4755	685	469225	26.1725
650	422500	25.4951	686	470596	26.1916
651	423801	25.5147	687	471969	26.2107
652	425104	25.5343	688	473344	26.2298
653	426409	25.5539	689	474721	26.2488
654	427716	25.5734	690	476100	26.2679
655	429025	25.5930	691	477481	26.2869
656	430336	25.6125	692	478864	26.3059
657	431649	25.6320	693	480249	26.3249
658	432964	25.6515	694	481636	26.3439
659	434281	25.6710	695	483025	26.3629
660	435600	25.6905	696	484416	26.3818
661	436921	25.7099	697	485809	26.4008
662	438244	25.7294	698	487204	26.4197
663	439569	25.7488	699	488601	26.4386
664	440896	25.7682	700	480000	26.4575
665	442225	25.7876	701	491401	26.4764
666	443556	25.8070	702	492804	26.4953
667	444889	25.8263	703	494209	26.5141
668	446224	25.8457	704	495616	26.5330
669	447561	25.8650	705	497025	26.5518
670	448900	25.8844	706	498436	26.5707
671	450241	25.9037	707	499849	26.5895
672	451584	25.9230	708	501264	26.6083
673	452929	25.9422	709	502681	26.6271
674	454276	25.9615	710	504100	26.6458
675	455625	25.9809	711	505521	26.6646
676	456976	26.0000	712	506944	26.6833

Squares and Square Roots of Numbers

No.	Square	Square Root	No.	Square	Square Root
713	508369	26.7021	749	561001	27.3679
714	509796	26.7208	750	562500	27.3861
715	511225	26.7395	751	564001	27.4044
716	512656	26.7582	752	565504	27.4226
717	514089	26.7769	753	567009	27.4408
718	515524	26.7955	754	568516	27.4591
719	516961	26.8142	755	570025	27.4773
720	518400	26.8328	756	571536	27.4955
721	519841	26.8514	757	573049	27.5136
722	521284	26.8701	758	574564	27.5318
723	522729	26.8887	759	576081	27.5500
724	524176	26.9072	760	577600	27.5681
725	525625	26.9258	761	579121	27.5862
726	527076	26.9444	762	580644	27.6043
727	528529	26.9629	763	582169	27.6225
728	529984	26.9815	764	583696	27.6405
729	531441	27.0000	765	585225	27.6586
730	532900	27.0185	766	586756	27.6767
731	534361	27.0370	767	588289	27.6948
732	535824	27.0555	768	589824	27.7128
733	537289	27.0740	769	591361	27.7308
734	538756	27.0924	770	592900	27.7489
735	540225	27.1109	771	594441	27.7669
736	541696	27.1293	772	595984	27.7849
737	543169	27.1477	773	597529	27.8029
738	544644	27.1662	774	599076	27.8209
739	546121	27.1846	775	600625	27.8388
740	547600	27.2029	776	602176	27.8568
741	549801	27.2213	777	603729	27.8747
742	550564	27.2397	778	605284	27.8927
743	552049	27.2580	779	606841	27.9106
744	553536	27.2764	780	608400	27.9285
745	555025	27.2947	781	609961	27.9464
746	556516	27.3130	782	611524	27.9643
747	558009	27.3313	783	613089	27.9821
748	559504	27.3496	784	614656	28.0000

Squares and Square Roots of Numbers

No.	Square	Square Root	No.	Square	Square Root
785	616225	28.0179	821	674041	28.6531
786	617796	28.0357	822	675684	28.6705
787	619369	28.0535	823	677329	28.6880
788	620944	28.0713	824	678976	28.7054
789	622521	28.0891	825	680625	28.7228
790	624100	28.1069	826	682276	28.7402
791	625681	28.1247	827	683929	28.7576
792	627264	28.1425	828	685584	28.7750
793	628849	28.1603	829	687241	28.7924
794	630436	28.1780	830	688900	28.8097
795	632025	28.1957	831	690561	28.8271
796	633616	28.2135	832	692224	28.8444
797	635209	28.2312	833	693889	28.8617
798	636804	28.2489	834	695556	28.8791
799	638401	28.2666	835	697225	28.8964
800	640000	28.2843	836	698896	28.9137
801	641601	28.3019	837	700659	28.9310
802	643204	28.3196	838	702244	28.9482
803	644809	28.3373	839	703921	28.9655
804	646416	28.3549	840	705600	28.9828
805	648025	28.3725	841	707281	29.0000
806	649636	28.3901	842	708964	29.0172
807	651249	28.4077	843	710649	29.0345
808	652864	28.4253	844	712336	29.0517
809	654481	28.4429	845	714025	29.0689
810	656100	28.4605	846	715716	29.0861
811	657721	28.4781	847	717409	29.1033
812	659344	28.4956	848	719104	29.1204
813	660969	28.5132	849	720801	29.1376
814	662596	28.5307	850	722500	29.1548
815	664225	28.5482	851	724201	29.1719
816	665856	28.5657	852	725904	29.1890
817	667489	28.5832	853	727609	29.2062
818	669124	28.6007	854	729316	29.2233
819	670761	28.6182	855	731025	29.2404
820	672400	28.6356	856	732736	29.2575

Squares and Square Roots of Numbers

No.	Square	Square Root	No.	Square	Square Root
857	734449	29.2746	893	797449	29.8831
858	736164	29.2916	894	799236	29.8998
859	737881	29.3087	895	801025	29.9166
860	739600	29.3258	896	802816	29.9333
861	741321	29.3428	897	804609	29.9500
862	743044	29.3598	898	806404	29.9666
863	744769	29.3769	899	808201	29.9833
864	746496	29.3939	900	810000	30.0000
865	748225	29.4109	901	811801	30.0167
866	749956	29.4279	902	813604	30.0333
867	751689	29.4449	903	815409	30.0500
868	753424	29.4618	904	817216	30.0666
869	755161	29.4788	905	819025	30.0832
870	756900	29.4958	906	820836	30.0998
871	758641	29.5127	907	822649	30.1164
872	760384	29.5296	908	824464	30.1330
873	762129	29.5466	909	826281	30.1496
874	763876	29.5635	910	828100	30.1662
875	765625	29.5804	911	829921	30.1828
876	767376	29.5973	912	731744	30.1993
877	769129	29.6142	913	833569	30.2159
878	770884	29.6311	914	835396	30.2324
879	772641	29.6479	915	837225	30.2490
880	774400	29.6648	916	839056	30.2655
881	776161	29.6816	917	840899	30.2820
882	777924	29.6985	918	842724	30.2985
883	779689	29.7153	919	844561	30.3150
884	781456	29.7321	920	846400	30.3315
885	783225	29.7489	921	848241	30.3480
886	784996	29.7658	922	850084	30.3645
887	786769	29.7825	823	851929	30.3809
888	788544	29.7993	924	853776	30.3974
889	790321	29.8161	925	855625	30.4138
890	792100	29.8329	926	857476	30.4302
891	793881	29.8496	927	859329	30.4467
892	795664	29.8664	928	861184	30.4631

Squares and Square Roots of Numbers

No.	Square	Square Root	No.	Square	Square Root
929	863041	30.4795	965	931225	31.0644
930	864900	30.4959	966	933156	31.0805
931	866761	30.5123	967	935089	31.0966
932	868624	30.5287	968	937024	31.1127
933	870489	30.5450	969	938961	31.1288
934	872356	30.5614	970	940900	31.1448
935	874225	30.5778	971	942841	31.1609
936	876096	30.5941	972	944784	31.1769
937	877969	30.6105	973	946729	31.1929
938	879844	30.6268	974	948676	31.2090
939	881721	30.6431	975	950625	31.2250
940	883600	30.6594	976	952576	31.2410
941	885481	30.6757	977	954529	31.2570
942	887364	30.6920	978	956484	31.2730
943	889249	30.7083	979	958441	31.2890
944	891136	30.7246	980	960400	31.3050
945	893025	30.7409	981	962361	31.3209
946	894916	30.7571	982	964324	31.3369
947	896809	30.7734	983	966289	31.3528
948	898704	30.7896	984	968256	31.3688
949	900601	30.8058	985	970225	31.3847
950	902500	30.8221	986	972196	31.4006
951	904401	30.8383	987	974169	31.4166
952	906304	30.8545	988	976144	31.4325
953	908209	30.8707	989	978121	31.4484
954	910116	30.8869	990	980100	31.4643
955	912025	30.9031	991	982081	31.4802
956	913936	30.9192	992	984064	31.4960
957	915849	30.9354	993	986049	31.5119
958	917764	30.9516	994	988036	31.5278
959	919681	30.9677	995	990025	31.5436
960	921600	30.9839	996	992016	31.5595
961	923521	31.0000	997	994009	31.5753
962	925444	31.0161	998	996004	31.5911
963	927369	31.0322	999	998001	31.6070
964	929296	31.0483	1000	1000000	31.6228

Areas and Volumes

Square	$A = $ area. $A = s^2$ $A = \frac{1}{2}\, d^2$ $s = 0.7071\, d = \sqrt{A}$ $d = 1.414\, s = 1.414\, \sqrt{A}$
Rectangle	$A = $ area. $A = ab$ $d = \sqrt{a^2 + b^2}$ $a = A \div b$ $b = A \div a$
Parallelogram	$A = $ area. $A = ab$ $a = A \div b$ $b = A \div a$ Note that dimension a is measured at right angles to line b.
Right-angled triangle	$A = $ area. $A = \dfrac{bc}{2}$ $A = \sqrt{b^2 + c^2}$ $b = \sqrt{a^2 - c^2}$ $c = \sqrt{a^2 - b^2}$
Acute-angled triangle	$A = $ area. $A = \dfrac{bh}{2} = \dfrac{b}{2}\sqrt{a^2 - \left(\dfrac{a^2 + b^2 - c^2}{2\,b}\right)^2}$
Obtuse-angled triangle	$A = $ area. $A = \dfrac{bh}{2} = \dfrac{b}{2}\sqrt{a^2 - \left(\dfrac{c^2 - a^2 - b^2}{2\,b}\right)^2}$

Areas and Volumes

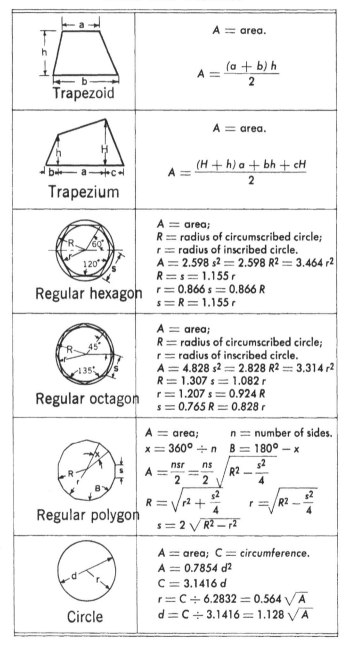

Trapezoid	A = area. $$A = \frac{(a + b)\, h}{2}$$
Trapezium	A = area. $$A = \frac{(H + h)\, a + bh + cH}{2}$$
Regular hexagon	A = area; R = radius of circumscribed circle; r = radius of inscribed circle. $A = 2.598\, s^2 = 2.598\, R^2 = 3.464\, r^2$ $R = s = 1.155\, r$ $r = 0.866\, s = 0.866\, R$ $s = R = 1.155\, r$
Regular octagon	A = area; R = radius of circumscribed circle; r = radius of inscribed circle. $A = 4.828\, s^2 = 2.828\, R^2 = 3.314\, r^2$ $R = 1.307\, s = 1.082\, r$ $r = 1.207\, s = 0.924\, R$ $s = 0.765\, R = 0.828\, r$
Regular polygon	A = area; n = number of sides. $x = 360° \div n$ $B = 180° - x$ $A = \dfrac{nsr}{2} = \dfrac{ns}{2}\sqrt{R^2 - \dfrac{s^2}{4}}$ $R = \sqrt{r^2 + \dfrac{s^2}{4}}$ $r = \sqrt{R^2 - \dfrac{s^2}{4}}$ $s = 2\sqrt{R^2 - r^2}$
Circle	A = area; C = circumference. $A = 0.7854\, d^2$ $C = 3.1416\, d$ $r = C \div 6.2832 = 0.564\,\sqrt{A}$ $d = C \div 3.1416 = 1.128\,\sqrt{A}$

Areas and Volumes

Circular sector	A = area; l = length of arc; x = angle, in degrees. $$l = \frac{2A}{r}$$ $A = \frac{1}{2}\,rl$ $$x = \frac{57.296\,l}{r} \qquad r = \frac{2A}{l}$$
Circular segment	A = area; l = length of arc; x = angle, in degrees. $$c = 2\sqrt{h\,(2r-h)}$$ $$A = \frac{1}{2}\,[rl - c\,(r-h)]$$ $$l = 0.01745\,rx$$ $$h = r - \frac{1}{2}\sqrt{4\,r^2 - c^2}$$ $$x = \frac{57.296\,l}{r}$$
Ellipse	A = area; P = perimeter or circumference. $A = 3.1416\,ab.$ An approximate formula for the perimeter is: $$P = 3.1416\sqrt{2\,(a^2 + b^2)}$$
Cube	V = volume. $$V = s^3$$
Square prism	V = volume. $$V = abc$$ $$a = \frac{V}{bc} \qquad b = \frac{V}{ac} \qquad c = \frac{V}{ab}$$
Prism	V = volume; A = area of end surface. $$V = h \times A$$ The area A of the end surface is found by the formulas for areas of plane figures on the preceding pages. Dimension h must be measured perpendicular to end surface.

Areas and Volumes

 Pyramid	$V = $ volume. $V = \frac{1}{3} h \times$ area of base.
 Frustum of pyramid	$V = $ volume. $V = \frac{h}{3}\left(A_1 + A_2 + \sqrt{A_1 \times A_2}\right)$
 Cylinder	$V = $ volume; $S = $ area of curved surface. $V = 0.7854\ d^2 h$ $S = 6.2832\ rh$ Area of all surfaces is S plus twice the area of an end.
 Portion of cylinder	$V = $ volume; $S = $ area of curved surface. $V = 0.3927\ d^2\ (h_1 + h_2)$ $S = 1.5708\ d\ (h_1 + h_2)$
 Hollow cylinder	$V = $ volume. $V = 0.7854\ h\ (D^2 - d^2)$ $= 3.1416\ ht\ (R + r)$
 Cone	$V = $ volume; $A = $ area of conical surface. $V = 0.2618\ d^2 h$ $A = 1.5708\ ds$ $s = \sqrt{r^2 + h^2}$

Areas and Volumes

Frustum of cone	V = volume; A = area of conical surface. $V = 1.0472\ h\ (R^2 + Rr + r^2)$ $A = 3.1416\ s\ (R + r)$ $a = R - r \qquad s = \sqrt{a^2 + h^2}$
Sphere	V = volume; A = area of surface. $V = 0.5236\ d^3$ $A = 3.1416\ d^2$ $r = 0.6204 \times$ cube root of V
Spherical sector	V = volume; A = total area of conical and spherical surface. $V = 2.0944\ r^2h$ $A = 3.1416\ r\ (2h + \frac{1}{2}\ c)$ $c = 2\ \sqrt{h\ (2r - h)}$
Spherical segment	V = volume; A = area of spherical surface. $V = 3.1416\ h^2 \left(r - \dfrac{h}{3} \right)$ $A = 2\pi rh$ $c = 2\ \sqrt{h\ (2r - h)}$
Spherical zone	V = volume; A = area of spherical surface. $V = 0.5236\ h \left(\dfrac{3\ c_1^2}{4} + \dfrac{3\ c_2^2}{4} + h^2 \right)$ $A = 2\pi rh = 6.2832\ rh$
Barrel	V = approximate volume. If sides are bent to the arc of a circle: $V = 0.262\ h\ (2\ D^2 + d^2)$ If sides are bent to the arc of a parabola: $V = 0.209\ h\ (2\ D^2 + Dd + \frac{3}{4}\ d^2)$

Reciprocals of Numbers from 1 to 99

No.	Reciprocal	No.	Reciprocal	No.	Reciprocal
1	1.00000000	34	.02941176	67	.01492537
2	.50000000	35	.02857143	68	.01470588
3	.33333333	36	.02777778	69	.01449275
4	.25000000	37	.02702703	70	.01428571
5	.20000000	38	.02631579	71	.01408451
6	.16666667	39	.02564103	72	.01388889
7	.14285714	40	.02500000	73	.01369863
8	.12500000	41	.02439024	74	.01351351
9	.11111111	42	.02380952	75	.01333333
10	.10000000	43	.02325581	76	.01315789
11	.09090909	44	.02272727	77	.01298701
12	.08333333	45	.02222222	78	.01282051
13	.07692308	46	.02173913	79	.01265823
14	.07142857	47	.02127660	80	.01250000
15	.06666667	48	.02083333	81	.01234568
16	.06250000	49	.02040816	82	.01219512
17	.05882353	50	.02000000	83	.01204819
18	.05555556	51	.01960784	84	.01190476
19	.05263158	52	.01923077	85	.01176471
20	.05000000	53	.01886792	86	.01162791
21	.04761905	54	.01851852	87	.01149425
22	.04545455	55	.01818182	88	.01136364
23	.04347826	56	.01785714	89	.01123595
24	.04166667	57	.01754386	90	.01111111
25	.04000000	58	.01724138	91	.01098901
26	.03846154	59	.01694915	92	.01086956
27	.03703704	60	.01666667	93	.01075269
28	.03571429	61	.01639344	94	.01063830
29	.03448276	62	.01612903	95	.01052632
30	.03333333	63	.01587302	96	.01041667
31	.03225806	64	.01562500	97	.01030928
32	.03125000	65	.01538461	98	.01020408
33	.03030303	66	.01515151	99	.01010101

Measures of Length

1 mile = 1760 yards = 5280 feet.

1 yard = 3 feet = 36 inches.   1 foot = 12 inches.

1 mil = 0.001 inch.   1 fathom = 2 yards = 6 feet.

1 rod = 5.5 yards = 16.5 feet.

1 hand = 4 inches.   1 span = 9 inches.

1 micro-inch = one millionth inch or 0.000001 inch.

(1 micron = one millionth meter = 0.00003937 inch.)

Surveyor's Measure

1 mile = 8 furlongs = 80 chains.

1 furlong = 10 chains = 220 yards.

1 chain = 4 rods = 22 yards = 66 feet = 100 links.

1 link = 7.92 inches.

Nautical Measure

1 league = 3 nautical miles.

1 nautical mile = 6080.2 feet = 1.1516 statute mile.

(The *knot,* which is nautical unit of speed, is equivalent to a speed of 1 nautical mile per hour.)

One degree at the equator = 60 nautical miles = 69.096 statute miles. 360 degrees = 21,600 nautical miles = 24,874.5 statute miles = circumference at equator.

Square Measure

1 square mile = 640 acres = 6400 square chains.

1 acre = 10 square chains = 4840 square yards = 43,560 square feet.

1 square chain = 16 square rods = 484 square yards = 4356 square feet.

1 square rod = 30.25 square yards = 272.25 square feet = 625 square links.

1 square yard = 9 square feet.

1 square foot = 144 square inches.

An acre is equal to a square, the side of which is 208.7 feet.

Measure Used for Diameters and Areas of Electric Wires

1 circular inch = area of circle 1 inch in diameter = 0.7854 square inch.

1 circular inch = 1,000,000 circular mils.

1 square inch = 1.2732 circular inch = 1,273,239 circular mils.

A circular mil is the area of a circle 0.001 inch in diameter.

Cubic Measure

1 cubic yard = 27 cubic feet.

1 cubic foot = 1728 cubic inches.

The following cubic measures are also used for wood and masonry:

1 cord of wood = 4 x 4 x 8 feet = 128 cubic feet.

1 perch of masonry = $16\frac{1}{2}$ x $1\frac{1}{2}$ x 1 foot = $24\frac{3}{4}$ cubic feet.

Shipping Measure

For measuring entire internal capacity of a vessel:

1 register ton = 100 cubic feet.

For measurement of cargo:

Approximately 40 cubic feet of merchandise is considered a shipping ton, unless that bulk would weigh more than 2000 pounds, in which case the freight charge may be based upon weight.

40 cubic feet = 32.143 U. S. bushels = 31.16 Imperial bushels.

Dry Measure

1 bushel (U. S. or Winchester struck bushel) = 1.2445 cubic foot = 2150.42 cubic inches.

1 bushel = 4 pecks = 32 quarts = 64 pints.

1 peck = 8 quarts = 16 pints.

1 quart = 2 pints.

1 heaped bushel = $1\frac{1}{4}$ struck bushel.

1 cubic foot = 0.8036 struck bushel.

1 British Imperial bushel = 8 Imperial gallons = 1.2837 cubic foot = 2218.19 cubic inches.

Liquid Measure

1 U. S. gallon = 0.1337 cubic foot = 231 cubic inches = 4 quarts = 8 pints.

1 quart = 2 pints = 8 gills.

1 pint = 4 gills.

1 British Imperial gallon = 1.2009 U. S. gallon = 277.42 cubic inches.

1 cubic foot = 7.48 U. S. gallons.

Apothecaries' Fluid Measure

1 U. S. fluid ounce = 8 drachms = 1.805 cubic inch = 1/128 U. S. gallon.

1 fluid drachm = 60 minims.

1 British fluid ounce = 1.732 cubic inch.

Old Liquid Measure

1 tun = 2 pipes = 3 puncheons.
1 pipe or butt = 2 hogsheads = 4 barrels = 126 gallons.
1 puncheon = 2 tierces = 84 gallons.
1 hogshead = 2 barrels = 63 gallons.
1 tierce = 42 gallons.
1 barrel = 31½ gallons.

Avoirdupois or Commercial Weight

1 gross or long ton = 2240 pounds.
1 net or short ton = 2000 pounds.
1 pound = 16 ounces = 7000 grains.
1 ounce = 16 drachms = 437.5 grains.
The following measures for weight are now seldom used in the United States:
1 hundred-weight = 4 quarters = 112 pounds (1 gross or long ton = 20 hundred-weights); 1 quarter = 28 pounds; 1 stone = 14 pounds; 1 quintal = 100 pounds.

Apothecaries' Weight

1 pound = 12 ounces = 5760 grains.
1 ounce = 8 drachms = 480 grains.
1 drachm = 3 scruples = 60 grains.
1 scruple = 20 grains.

Measures of Pressure

1 pound per square inch = 144 pounds per square foot = 0.068 atmosphere = 2.042 inches of mercury at 62 degrees F. = 27.7 inches of water at 62 degrees F. = 2.31 feet of water at 62 degrees F.
1 atmosphere = 30 inches of mercury at 62 degrees F. = 14.7 pounds per square inch = 2116.3 pounds per square foot = 33.95 feet of water at 62 degrees F.
1 foot of water at 62 degrees F. = 62.355 pounds per square foot = 0.433 pound per square inch.
1 inch of mercury at 62 degrees F. = 1.132 foot of water = 13.58 inches of water = 0.491 pound per square inch.

Inches Converted to Decimals
of a Foot

Inches	Decimal of a Foot	Inches	Decimal of a Foot	Inches	Decimal of a Foot
⅛	.0104	3⅛	.2604	6¼	.5208
¼	.0208	3¼	.2708	6½	.5417
⅜	.0313	3⅜	.2813	6¾	.5625
½	.0417	3½	.2917	7	.5833
⅝	.0521	3⅝	.3021	7¼	.6042
¾	.0625	3¾	.3125	7½	.6250
⅞	.0729	3⅞	.3229	7¾	.6458
1	.0833	4	.3333	8	.6667
1⅛	.0938	4⅛	.3438	8¼	.6875
1¼	.1042	4¼	.3542	8½	.7083
1⅜	.1146	4⅜	.3646	8¾	.7292
1½	.1250	4½	.3750	9	.7500
1⅝	.1354	4⅝	.3854	9¼	.7708
1¾	.1458	4¾	.3958	9½	.7917
1⅞	.1563	4⅞	.4063	9¾	.8125
2	.1667	5	.4167	10	.8333
2⅛	.1771	5⅛	.4271	10¼	.8542
2¼	.1875	5¼	.4375	10½	.8750
2⅜	.1979	5⅜	.4479	10¾	.8958
2½	.2083	5½	.4583	11	.9167
2⅝	.2188	5⅝	.4688	11¼	.9375
2¾	.2292	5¾	.4792	11½	.9583
2⅞	.2396	5⅞	.4896	11¾	.9792
3	.2500	6	.5000	12	1.0000

Example: 4⅜ in. is 0.36458 of a foot.

Fahrenheit-Centigrade

Conversion

In the temperature conversion tables on following pages, numbers in the center column, in bold face type, refer to the temperature in either Fahrenheit or Centigrade degrees. If it is desired to convert from Fahrenheit to Centigrade degrees, consider the center column as a table of Fahrenheit temperatures and read the corresponding Centigrade temperature in the column at the left. If it is desired to convert from Centigrade to Fahrenheit degrees, consider the center column as a table of Centigrade values, and read the corresponding Fahrenheit temperature on the right.

For example, if it is desired to convert 40 degrees Fahrenheit to Centigrade degrees, the column at left shows 4.4 degrees Centigrade for a reading of 40 (degrees Fahrenheit) in the center column.

Interpolation factors are given for use with that portion of the table in which the center column advances in increments of 10. To illustrate, suppose it is desired to find the Fahrenheit equivalent of 314 deg C. The equivalent of 310 deg C, found in the body of the main table, is seen to be 590 deg F. The Fahrenheit equivalent of a 4-deg C difference is seen to be 7.2, as read in the table of interpolating factors. The answer is the sum of the two, or 597.2 deg F.

For conversions not covered in the table, the following formulas are used:

$$F = 1.8 \ C + 32$$
$$C = (F-32) \div 1.8$$

Fahrenheit-Centigrade Conversion

Deg C		Deg F	Deg C		Deg F
—17.8	0	32—	0.6	33	91.4
—17.2	1	33.8	1.1	34	93.2
—16.7	2	35.6	1.7	35	95.0
—16.1	3	37.4	2.2	36	96.8
—15.6	4	39.2	2.7	37	98.6
—15.0	5	41.0	3.3	38	100.4
—14.4	6	42.8	3.9	39	102.2
—13.9	7	44.6	4.4	40	104.0
—13.3	8	46.4	5.0	41	105.8
—12.8	9	48.2	5.6	42	107.6
—12.2	10	50.0	6.1	43	109.4
—11.7	11	51.8	6.7	44	111.2
—11.1	12	53.6	7.2	45	113.0
—10.6	13	55.4	7.8	46	114.8
—10.0	14	57.2	8.3	47	116.6
— 9.4	15	59.0	8.9	48	118.4
— 8.9	16	60.8	9.4	49	120.2
— 8.3	17	62.6	10.0	50	122.0
— 7.8	18	64.4	10.6	51	123.8
— 7.2	19	66.2	11.1	52	125.6
— 6.7	20	68.0	11.7	53	127.4
— 6.1	21	69.8	12.2	54	129.2
— 5.6	22	71.6	12.8	55	131.0
— 5.0	23	73.4	13.3	56	132.8
— 4.4	24	75.2	13.9	57	134.6
— 3.9	25	77.0	14.4	58	136.4
— 3.3	26	78.8	15.0	59	138.2
— 2.8	27	80.6	15.6	60	140.0
— 2.2	28	82.4	16.1	61	141.8
— 1.7	29	84.2	16.7	62	143.6
— 1.1	30	86.0	17.2	63	145.4
— 0.6	31	87.8	17.8	64	147.2
0—	32	89.6	18.3	65	149.0

Fahrenheit-Centigrade Conversion

Deg C		Deg F	Deg C		Deg F
18.9	66	150.8	37.2	99	210.2
19.4	67	152.6	37.8	100	212.0
20.0	68	154.4	38.3	101	213.8
20.6	69	156.2	38.9	102	215.6
21.1	70	158.0	39.4	103	217.4
21.7	71	159.8	40.0	104	219.2
22.2	72	161.6	40.6	105	221.0
22.8	73	163.4	41.1	106	222.8
23.3	74	165.2	41.7	107	224.6
23.9	75	167.0	42.2	108	226.4
24.4	76	168.8	42.8	109	228.2
25.0	77	170.6	43.3	110	230.0
25.6	78	172.4	43.9	111	231.8
26.1	79	174.2	44.4	112	233.6
26.7	80	176.0	45.0	113	235.4
27.2	81	177.8	45.6	114	237.2
27.8	82	179.6	46.1	115	239.0
28.3	83	181.4	46.7	116	240.8
28.9	84	183.2	47.2	117	242.6
29.4	85	185.0	47.8	118	244.4
30.0	86	186.8	48.3	119	246.2
30.6	87	188.6	48.9	120	248.0
31.1	88	190.4	49.4	121	249.8
31.7	89	192.2	50.0	122	251.6
32.2	90	194.0	50.6	123	253.4
32.8	91	195.8	51.1	124	255.2
33.3	92	197.6	51.7	125	257.0
33.9	93	199.4	52.2	126	258.8
34.4	94	201.2	52.8	127	260.6
35.0	95	203.0	53.3	128	262.4
35.6	96	204.8	53.9	129	264.2
36.1	97	206.6	54.4	130	266.0
36.7	98	208.4	55.0	131	267.8

Fahrenheit-Centigrade Conversion

Deg C		Deg F	Deg C		Deg F
55.6	132	269.6	73.9	165	329.0
56.1	133	271.4	74.4	166	330.8
56.7	134	273.2	75.0	167	332.6
57.2	135	275.0	75.6	168	334.4
57.8	136	276.8	76.1	169	336.2
58.3	137	278.6	76.7	170	338.0
58.9	138	280.4	77.2	171	339.8
59.4	139	282.2	77.8	172	341.6
60.0	140	284.0	78.3	173	343.4
60.6	141	285.8	78.9	174	345.2
61.1	142	287.6	79.4	175	347.0
61.7	143	289.4	80.0	176	348.8
62.2	144	291.2	80.6	177	350.6
62.8	145	293.0	81.1	178	352.4
63.3	146	294.8	81.7	179	354.2
63.9	147	296.6	82.2	180	356.0
64.4	148	298.4	82.8	181	357.8
65.0	149	300.2	83.3	182	359.6
65.6	150	302.0	83.9	183	361.4
66.1	151	303.8	84.4	184	363.2
66.7	152	305.6	85.0	185	365.0
67.2	153	307.4	85.6	186	366.8
67.8	154	309.2	86.1	187	368.6
68.3	155	311.0	86.7	188	370.4
68.9	156	312.8	87.2	189	372.2
69.4	157	314.6	87.8	190	374.0
70.0	158	316.4	88.3	191	375.8
70.6	159	318.2	88.9	192	377.6
71.1	160	320.0	89.4	193	379.4
71.7	161	321.8	90.0	194	381.2
72.2	162	323.6	90.6	195	383.0
72.8	163	325.4	91.1	196	384.8
73.3	164	327.2	91.7	197	386.6

Fahrenheit-Centigrade Conversion

Deg C		Deg F	Deg C		Deg F
92.2	198	388.4	204.4	400	752.0
92.8	199	390.2	210	410	770.0
93.3	200	392.0	215.6	420	788
93.9	201	393.8	221.1	430	806
94.4	202	395.6	226.7	440	824
95.0	203	397.4	232.2	450	842
95.6	204	399.2	237.8	460	860
96.1	205	401.0	243.3	470	878
96.7	206	402.8	248.9	480	896
97.2	207	404.6	254.4	490	914
97.8	208	406.4	260.0	500	932
98.3	209	408.2	265.6	510	950
98.9	210	410.0	271.1	520	968
99.4	211	411.8	276.7	530	986
100.0	212	413.6	282.2	540	1004
104.4	220	428.0	287.8	550	1022
110.0	230	446.0	293.3	560	1040
115.6	240	464.0	298.9	570	1058
121.1	250	482.0	304.4	580	1076
126.7	260	500.0	310.0	590	1094
132.2	270	518.0	315.6	600	1112
137.8	280	536.0	321.1	610	1130
143.3	290	554.0	326.7	620	1148
148.9	300	572.0	332.2	630	1166
154.4	310	590.0	337.8	640	1184
160.0	320	608.0	343.3	650	1202
165.6	330	626.0	348.9	660	1220
171.1	340	644.0	354.4	670	1238
176.7	350	662.0	360.0	680	1256
182.2	360	680.0	365.6	690	1274
187.8	370	698.0	371.1	700	1292
193.3	380	716.0	376.7	710	1310
198.9	390	734.0	382.2	720	1328

Conversion of Head in Feet of Water to Pounds per Square Inch

Head in Feet	Pounds per Sq. In.	Head in Feet	Pounds per Sq. In.
1	.43	140	60.63
2	.87	150	64.96
3	1.30	160	69.29
4	1.73	170	73.63
5	2.17	180	77.96
6	2.60	190	82.29
7	3.03	200	86.62
8	3.46	225	97.45
9	3.90	250	108.27
10	4.33	275	119.10
20	8.66	300	129.93
30	12.99	325	140.75
40	17.32	350	151.58
50	21.65	400	173.24
60	25.99	500	216.55
70	30.32	600	259.85
80	34.65	700	303.16
90	38.98	800	346.47
100	43.31	900	389.78
110	47.65	1000	433.09
120	51.97	1500	649.64
130	56.30	2000	866.18

Conversion of Pounds per Square Inch to Head in Feet of Water

Pounds per Sq. In.	Head in Feet	Pounds per Sq. In.	Head in Feet
1	2.31	120	277.07
2	4.62	125	288.62
3	6.93	130	300.16
4	9.24	140	323.25
5	11.54	150	346.34
6	13.85	160	369.43
7	16.16	170	392.52
8	18.47	180	415.61
9	20.78	190	438.90
10	23.09	200	461.78
15	34.63	225	519.51
20	46.18	250	577.24
25	57.72	275	643.03
30	69.27	300	692.69
40	92.36	325	750.41
50	115.45	350	808.13
60	138.54	375	865.89
70	161.63	400	922.58
80	184.72	500	1154.48
90	207.81	1000	2309.00
100	230.90	1500	3463.48
110	253.98	2000	4618.00

Cubic Feet into U.S. Gallons

Cubic Feet	Gallons	Cubic Feet	Gallons
0.1	0.75	30	224.4
0.2	1.50	40	299.2
0.3	2.24	50	374.0
0.4	2.99	60	448.8
0.5	3.74	70	523.6
0.6	4.49	80	598.4
0.7	5.24	90	673.2
0.8	5.98	100	748.1
0.9	6.73	200	1496.1
1.0	7.48	300	2244.2
2.0	14.96	400	2992.2
3.0	22.44	500	3740.3
4.0	29.92	600	4488.3
5.0	37.40	700	5236.4
6.0	44.88	800	5984.4
7.0	52.36	900	6732.5
8.0	59.84	1000	7480.5
9.0	67.32	5000	37402.6
10.0	74.81	10000	74805.2
20.0	149.61	50000	374025.9

EXAMPLE

Convert 555.5 cubic feet into U.S. gallons.

SOLUTION

From table above, find and add conversion figures for 500, 50, 5, and 0.5 cubic feet. Thus, 3740.3 + 374.0 + 37.4 + 3.74 = 4155.44 U.S. gallons.

U.S. Gallons into Cubic Feet

Gallons	Cubic Feet	Gallons	Cubic Feet
1	0.134	300	40.10
2	0.267	400	53.47
3	0.401	500	66.84
4	0.535	600	80.21
5	0.668	700	93.58
6	0.802	800	106.94
7	0.936	900	120.31
8	1.069	1000	133.68
9	1.203	2000	267.36
10	1.337	3000	401.04
20	2.674	4000	534.72
30	4.010	5000	668.40
40	5.347	6000	802.08
50	6.684	7000	935.76
60	8.021	8000	1069.44
70	9.358	9000	1203.12
80	10.694	10000	1336.81
90	12.031	50000	6684.03
100	13.368	100000	13368.06
200	26.736	500000	66840.28

EXAMPLE

Convert 4321 U.S. gallons into cubic feet.

SOLUTION

From table above, find and add conversion figures for 4000, 300, 20, and 1 U.S. gallons. Thus, 534.72 + 40.10 + 2.67 + 0.13 = 577.62 cu. ft.

Conversion Factors

Multiply	By	To Obtain
Atmosphere	29.92	Inches of mercury
Atmosphere	33.90	Feet of water
Atmosphere	14.70	Pounds per square in.
Barrels (oil)	42	Gallons
Boiler horsepower	33,475	Btu per hour
Btu	0.252	Calories
Centimeters	0.3937	Inches
Cubic feet	1728	Cubic inches
Cubic feet of water	7.48	Gallons of water
Cubic feet of water	62.37	Pounds of water
Cubic feet per minute	0.1247	Gallons per second
Cubic meters	35.314	Cubic feet
Cubic meters	264.2	U.S. gallons
Feet	0.3048	Meters
Feet of water	0.881	Inches of mercury
Feet of water	62.37	Pounds per square ft.
Feet of water	0.4335	Pounds per square in.
Feet of water	0.0295	Atmospheres
Feet per minute	0.01136	Miles per hour
Feet per minute	0.01667	Feet per second
Gallons (U.S.)	0.1337	Cubic feet
Gallons (U.S.)	231	Cubic inches
Gallons (U.S.)	8.3453	Pounds of water
Gallons (Imperial)	277.3	Cubic inches
Gallons (Imperial)	10	Pounds of water
Grams	0.03527	Ounces
Horsepower	33,000	Foot-pounds per min.
Horsepower	0.7457	Kilowatts
Horsepower (boiler)	33,471.9	Btu per hour

Conversion Factors

Multiply	By	To Obtain
Inches	2.54	Centimeters
Inches	25.4	Millimeters
Inches of mercury	1.131	Feet of water
Inches of mercury	0.4912	Pounds per square in.
Inches of water	0.03613	Pounds per square in.
Inches of water	5.202	Pounds per square ft.
Kilograms	2.2046	Pounds
Kilometers	0.6214	Miles
Kilowatts	1.341	Horsepower
Kilowatt-hours	3415	Btu
Liters	0.2642	U.S. gallons
Meters	39.37	Inches
Meters	3.2808	Feet
Miles	1.609	Kilometers
Millimeters	0.03937	Inches
Ounces	28.35	Grams
Pounds	0.4536	Kilograms
Pounds	7000	Grains
Pounds of water	0.01602	Cubic feet
Pounds of water	27.68	Cubic inches
Pounds of water	0.1198	Gallons
Pounds per square in.	2.309	Feet of water
Pounds per square in.	2.0416	Inches of mercury
Tons of refrigeration	12,000	Btu per hour
Tons (long)	2240	Pounds
Tons, metric	2204.6	Pounds
Tons (short)	2000	Pounds
Watts	.5692	Btu per minute
Watt-hours	3.415	Btu

Useful Formulas

Water. Unless otherwise specified, data on specific weight and volume of water are usually at 60 deg. F., although sometimes they are at 39.2 deg., at which point water has its greatest density. At 32 deg., water weighs 62.42 lb. per cu. ft.; at 60 deg., 62.37; at 100 deg., 62.00; at 200 deg., 60.13.

Steam. The latent heat of water is 970 Btu per lb. at atmospheric pressure. That is, after water is raised to 212 deg., 970 more Btu must be added per pound to convert the water to steam at 212 deg.

Btu (British thermal unit) is the amount of heat required to raise the temperature of one pound of water one degree F. in temperature at about 60 deg.

Temperature and Thermometers. The two principal thermometer scales are the Fahrenheit (used in the U.S.A.) and the Centigrade (used in most non-English speaking countries and in most scientific work). In the former (F) the freezing point is at 32, the boiling point at 212; so that there are 180 divisions or degrees between. With the Centigrade (C) thermometer, freezing is at zero, boiling at 100, with 100 divisions between.

To convert Centigrade degrees to Fahrenheit, multiply the C reading by 9, divide by 5, then add 32.

Example. What is 90 deg. C in Fahrenheit degrees? Multiply 90 by 9 and get 810. Divide by 5 and get 162. Add 32 and get 194 deg. F.

To convert Fahrenheit degrees to Centigrade, subtract 32, multiply by 5 and divide by 9.

Example. What is 104 deg. F in Centigrade degrees? Subtract 32 from 104 and obtain 72. Multiply by 5 and get 360. Divide by 9 and the answer is 40 deg. C.

Horsepower. One horsepower is equivalent to 33,000 foot-pounds per minute. A foot pound is the energy needed to raise 1 pound 1 foot vertically, or ½ pound 2 feet and so on. The power needed to produce 1 horsepower is thus equivalent to raising 33,000 lb. 1 foot, 3,300 lb. 10 feet, 33 lb. 1,000 feet *in one minute.* To determine the theoretical horsepower required to raise water a given height, multiply the gallons to be raised per minute by 8.33 to find the pounds to be raised per minute. This multiplied by the vertical distance between the source and the point to which the water is to be raised gives the theoretical horsepower.

Tanks. To find the capacity of a cylindrical tank square the diameter in feet, multiply by the length in feet and then by .7854. Result is capacity in cubic feet.

CAUTION. Due to lack of barrel standardization, the number of gallons in a barrel varies for many different liquids. Consequently, it is safer to use cubic feet as a measure rather than barrels.

EXAMPLE. What is the capacity of an 8 foot diameter 20 foot long cylindrical tank?

SOLUTION. 8 squared is 64. 64 X 20 = 1280 and 1280 multiplied by .7854 = 1005.3 cu. ft.

Trigonometric Functions

Degrees	Sine	Cosine	Tangent	Cotangent	Secant	Cosecant
0	0	1.00000	0	Inf.	1.0000	Inf.
½	.00873	.99996	.00873	114.59	1.0000	114.59
1	.01745	.99985	.01745	57.290	1.0001	57.299
1½	.02618	.99966	.02618	38.188	1.0003	38.201
2	.03490	.99939	.03492	28.636	1.0006	28.654
2½	.04362	.99905	.04366	22.904	1.0009	22.925
3	.05234	.99863	.05241	19.081	1.0014	19.107
3½	.06105	.99813	.06116	16.350	1.0019	16.380
4	.06976	.99756	.06993	14.301	1.0024	14.335
4½	.07846	.99692	.07870	12.706	1.0031	12.745
5	.08715	.99619	.08749	11.430	1.0038	11.474
5½	.09584	.99540	.09629	10.385	1.0046	10.433
6	.01453	.99452	.10510	9.5144	1.0055	9.5668
6½	.11320	.99357	.11393	8.7769	1.0065	8.8337
7	.12187	.99255	.12278	8.1443	1.0075	8.2055
7½	.13053	.99144	.13165	7.5957	1.0086	7.6613
8	.13917	.99027	.14054	7.1154	1.0098	7.1853
8½	.14781	.98901	.14945	6.6911	1.0111	6.7655
9	.15643	.98769	.15838	6.3137	1.0125	6.3924
9½	.16505	.98628	.16734	5.9758	1.0139	6.0588
10	.17365	.98481	.17633	5.6713	1.0154	5.7588
10½	.18223	.98325	.18534	5.3955	1.0170	5.4874
11	.19081	.98163	.19438	5.1445	1.0187	5.2408
11½	.19937	.97992	.20345	4.9151	1.0205	5.0158
12	.20791	.97815	.21256	4.7046	1.0223	4.8097
12½	.21644	.97630	.22169	4.5107	1.0243	4.6201
13	.22495	.97437	.23087	4.3315	1.0263	4.4454
13½	.23344	.97237	.24008	4.1653	1.0284	4.2836
14	.24192	.97029	.24933	4.0108	1.0306	4.1336
14½	.25038	.96815	.25862	3.8667	1.0329	3.9939
15	.25882	.96592	.26795	3.7320	1.0353	3.8637
15½	.26724	.96363	.27732	3.6059	1.0377	3.7420
16	.27564	.96126	.28674	3.4874	1.0403	3.6279
16½	.28401	.95882	.29621	3.3759	1.0429	3.5209
17	.29237	.95630	.30573	3.2708	1.0457	3.4203
17½	.30070	.95372	.31530	3.1716	1.0485	3.3255
18	.30902	.95106	.32492	3.0777	1.0515	3.2361
18½	.31730	.94832	.33459	2.9887	1.0545	3.1515

Trigonometric Functions

Degrees	Sine	Cosine	Tangent	Cotangent	Secant	Cosecant
19	.32557	.94552	.34433	2.9042	1.0576	3.0715
19½	.33381	.94264	.35412	2.8239	1.0608	2.9957
20	.34202	.93969	.36397	2.7475	1.0642	2.9238
20½	.35021	.93667	.37388	2.6746	1.0676	2.8554
21	.35837	.93358	.38386	2.6051	1.0711	2.7904
21½	.36650	.93042	.39391	2.5386	1.0748	2.7285
22	.37461	.92718	.40403	2.4751	1.0785	2.6695
22½	.38268	.92388	.41421	2.4142	1.0824	2.6131
23	.39073	.92050	.42447	2.3558	1.0864	2.5593
23½	.39875	.91706	.43481	2.2998	1.0904	2.5078
24	.40674	.91354	.44523	2.2460	1.0946	2.4586
24½	.41469	.90996	.45573	2.1943	1.0989	2.4114
25	.42262	.90631	.46631	2.1445	1.1034	2.3662
25½	.43051	.90258	.47697	2.0965	1.1079	2.3228
26	.43837	.89879	.48773	2.0503	1.1126	2.2812
26½	.44620	.89493	.49858	2.0057	1.1174	2.2411
27	.45399	.89101	.50952	1.9626	1.1223	2.2027
27½	.46175	.88701	.52057	1.9210	1.1274	2.1657
28	.46947	.88295	.53171	1.8807	1.1326	2.1300
28½	.47716	.87882	.54295	1.8418	1.1379	2.0957
29	.48481	.87462	.55431	1.8040	1.1433	2.0627
29½	.49242	.87035	.56577	1.7675	1.1489	2.0308
30	.50000	.86603	.57735	1.7320	1.1547	2.0000
30½	.50754	.86163	.58904	1.6977	1.1606	1.9703
31	.51504	.85717	.60086	1.6643	1.1666	1.9416
31½	.52250	.85264	.61280	1.6318	1.1728	1.9139
32	.52992	.84805	.62487	1.6003	1.1792	1.8871
32½	.53730	.84339	.63707	1.5697	1.1857	1.8611
33	.54464	.83867	.64941	1.5399	1.1924	1.8361
33½	.55194	.83388	.66188	1.5108	1.1992	1.8118
34	.55919	.82904	.67451	1.4826	1.2062	1.7883
34½	.56641	.82413	.68728	1.4550	1.2134	1.7655
35	.57358	.81915	.70021	1.4281	1.2208	1.7434
35½	.58070	.81411	.71329	1.4019	1.2283	1.7220
36	.58778	.80902	.72654	1.3764	1.2361	1.7013
36½	.59482	.80386	.73996	1.3514	1.2440	1.6812

Trigonometric Functions

Degrees	Sine	Cosine	Tangent	Cotangent	Secant	Cosecant
37	.60181	.79863	.75355	1.3270	1.2521	1.6616
37½	.60876	.79335	.76733	1.3032	1.2605	1.6427
38	.61566	.78801	.78128	1.2799	1.2690	1.6243
38½	.62251	.78261	.79543	1.2572	1.2778	1.6064
39	.62932	.77715	.80978	1.2349	1.2867	1.5890
39½	.63608	.77162	.82434	1.2131	1.2960	1.5721
40	.64279	.76604	.83910	1.1917	1.3054	1.5557
40½	.64945	.76041	.85408	1.1708	1.3151	1.5398
41	.65606	.75471	.86929	1.1504	1.3250	1.5242
41½	.66262	.74895	.88472	1.1303	1.3352	1.5092
42	.66913	.74314	.90040	1.1106	1.3456	1.4945
42½	.67559	.73728	.91633	1.0913	1.3563	1.4802
43	.68200	.73135	.93251	1.0724	1.3673	1.4663
43½	.68835	.72537	.94896	1.0538	1.3786	1.4527
44	.69466	.71934	.96569	1.0355	1.3902	1.4395
44½	.70091	.71325	.98270	1.0176	1.4020	1.4267
45	.70711	.70711	1.00000	1.0000	1.4142	1.4142
45½	.71325	.70091	1.0176	.98270	1.4267	1.4020
46	.71934	.69466	1.0355	.96569	1.4395	1.3902
46½	.72357	.68835	1.0538	.94896	1.4527	1.3786
47	.73135	.68200	1.0724	.93251	1.4663	1.3673
47½	.73728	.67559	1.0913	.91633	1.4802	1.3563
48	.74314	.66913	1.1106	.90040	1.4945	1.3456
48½	.74895	.66262	1.1303	.88412	1.5092	1.3352
49	.75471	.65606	1.1504	.86929	1.5242	1.3250
49½	.76041	.64945	1.1708	.85408	1.5398	1.3151
50	.76604	.64279	1.1917	.83910	1.5557	1.3054
50½	.77162	.63608	1.2131	.82434	1.5721	1.2960
51	.77715	.62932	1.2349	.80978	1.5890	1.2867
51½	.78261	.62251	1.2572	.79543	1.6064	1.2778
52	.78801	.61566	1.2799	.78128	1.6243	1.2690
52½	.79335	.60876	1.3032	.76733	1.6427	1.2605
53	.79863	.60181	1.3270	.75355	1.6616	1.2521
53½	.80386	.59482	1.3514	.73996	1.6812	1.2440
54	.80902	.58778	1.3764	.72654	1.7013	1.2361
54½	.81411	.58070	1.4019	.71329	1.7220	1.2283

Trigonometric Functions

Degrees	Sine	Cosine	Tangent	Cotangent	Secant	Cosecant
55	.81915	.57358	1.4281	.70021	1.7434	1.2208
55½	.82413	.56641	1.4550	.68728	1.7655	1.2134
56	.82904	.55919	1.4826	.67451	1.7883	1.2062
56½	.83388	.55194	1.5108	.66188	1.8118	1.1992
57	.83867	.54464	1.5399	.64941	1.8361	1.1924
57½	.84339	.53730	1.5697	.63707	1.8611	1.1857
58	.84805	.52992	1.6003	.62487	1.8871	1.1792
58½	.85264	.52250	1.6318	.61280	1.9139	1.1728
59	.85717	.51504	1.6643	.60086	1.9416	1.1666
59½	.86163	.50754	1.6977	.58904	1.9703	1.1606
60	.86603	.50000	1.7320	.57735	2.0000	1.1547
60½	.87035	.49242	1.7675	.56577	2.0308	1.1489
61	.87462	.48481	1.8040	.55431	2.0627	1.1433
61½	.87882	.47716	1.8418	.54295	2.0957	1.1379
62	.88295	.46947	1.8807	.53171	2.1300	1.1326
62½	.88701	.46175	1.9210	.52057	2.1657	1.1274
63	.89101	.45399	1.9626	.50952	2.2027	1.1223
63½	.89493	.44620	2.0057	.49858	2.2411	1.1174
64	.89879	.43837	2.0503	.48773	2.2812	1.1126
64½	.90258	.43051	2.0965	.47697	2.3228	1.1079
65	.90631	.42262	2.1445	.46631	2.3662	1.1034
65½	.90996	.41469	2.1943	.45573	2.4114	1.0989
66	.91354	.40674	2.2460	.44523	2.4586	1.0946
66½	.91706	.39875	2.2998	.43481	2.5078	1.0904
67	.92050	.39073	2.3558	.42447	2.5593	1.0864
67½	.92388	.38268	2.4142	.41421	2.6131	1.0824
68	.92718	.37461	2.4751	.40403	2.6695	1.0785
68½	.93042	.36650	2.5386	.39391	2.7285	1.0748
69	.93358	.35837	2.6051	.38386	2.7904	1.0711
69½	.93667	.35021	2.6746	.37388	2.8554	1.0676
70	.93969	.34202	2.7475	.36397	2.9238	1.0642
70½	.94264	.33381	2.8239	.35412	2.9957	1.0608
71	.94552	.32557	2.9042	.34433	3.0715	1.0576
71½	.94832	.31730	2.9887	.33459	3.1515	1.0545
72	.95106	.30902	3.0777	.32492	3.2361	1.0515
72½	.95372	.30070	3.1716	.31530	3.3255	1.0485

Trigonometric Functions

Degrees	Sine	Cosine	Tangent	Cotangent	Secant	Cosecant
73	.95630	.29237	3.2708	.30573	3.4203	1.0457
73½	.95882	.28401	3.3759	.29621	3.5209	1.0429
74	.96126	.27564	3.4874	.28674	3.6279	1.0403
74½	.96363	.26724	3.6059	.27732	3.7420	1.0377
75	.96592	.25882	3.7320	.26795	3.8637	1.0353
75½	.96815	.25038	3.8667	.25862	3.9939	1.0329
76	.97029	.24192	4.0108	.24933	4.1336	1.0306
76½	.97237	.23344	4.1653	.24008	4.2836	1.0284
77	.97437	.22495	4.3315	.23087	4.4454	1.0263
77½	.97630	.21644	4.5107	.22169	4.6201	1.0243
78	.97815	.20791	4.7046	.21256	4.8097	1.0223
78½	.97992	.19937	4.9151	.20345	5.0158	1.0205
79	.98163	.19081	5.1445	.19438	5.2408	1.0187
79½	.98325	.18223	5.3955	.18534	5.4874	1.0170
80	.98481	.17365	5.6713	.17633	5.7588	1.0154
80½	.98628	.16505	5.9758	.16734	6.0588	1.0139
81	.98769	.15643	6.3137	.15838	6.3924	1.0125
81½	.98901	.14781	6.6911	.14945	6.7655	1.0111
82	.99027	.13917	7.1154	.14054	7.1853	1.0098
82½	.99144	.13053	7.5957	.13165	7.6613	1.0086
83	.99255	.12187	8.1443	.12278	8.2055	1.0075
83½	.99357	.11320	8.7769	.11393	8.8337	1.0065
84	.99452	.10453	9.5144	.10510	9.5668	1.0055
84½	.99540	.09584	10.385	.09629	10.433	1.0046
85	.99619	.08715	11.430	.08749	11.474	1.0038
85½	.99692	.07846	12.706	.07870	12.745	1.0031
86	.99756	.06976	14.301	.06993	14.335	1.0024
86½	.99813	.06105	16.350	.06116	16.380	1.0019
87	.99863	.05234	19.081	.05241	19.107	1.0014
87½	.99905	.04362	22.904	.04366	22.925	1.0009
88	.99939	.03490	28.636	.03492	28.654	1.0006
88½	.99966	.02618	38.188	.02618	38.201	1.0003
89	.99985	.01745	57.290	.01745	57.299	1.0001
89½	.99996	.00873	114.59	.00873	114.59	1.0000
90	1.00000	0	Inf.	0	Inf.	1.0000

Solution of Right-angled Triangles

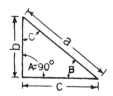

As shown in the illustration, the sides of the right-angled triangle are designated a, b and c. The angles opposite each of these sides are designated A, B and C, respectively.

Angle A, opposite the hypotenuse a is the right angle, and is therefore always one of the known quantities.

Sides and Angles Known	Formulas for Sides and Angles to be Found		
Sides a and b	$c = \sqrt{a^2 - b^2}$	$\sin B = \dfrac{b}{a}$	$C = 90° - B$
Sides a and c	$b = \sqrt{a^2 - c^2}$	$\sin C = \dfrac{c}{a}$	$B = 90° - C$
Sides b and c	$a = \sqrt{b^2 + c^2}$	$\tan B = \dfrac{b}{c}$	$C = 90° - B$
Side a; angle B	$b = a \times \sin B$	$c = a \times \cos B$	$C = 90° - B$
Side a; angle C	$b = a \times \cos C$	$c = a \times \sin C$	$B = 90° - C$
Side b; angle B	$a = \dfrac{b}{\sin B}$	$c = b \times \cot B$	$C = 90° - B$
Side b; angle C	$a = \dfrac{b}{\cos C}$	$c = b \times \tan C$	$B = 90° - C$
Side c; angle B	$a = \dfrac{c}{\cos B}$	$b = c \times \tan B$	$C = 90° - B$
Side c; angle C	$a = \dfrac{c}{\sin C}$	$b = c \times \cot C$	$B = 90° - C$

Examples of Solution of Right-angled Triangles

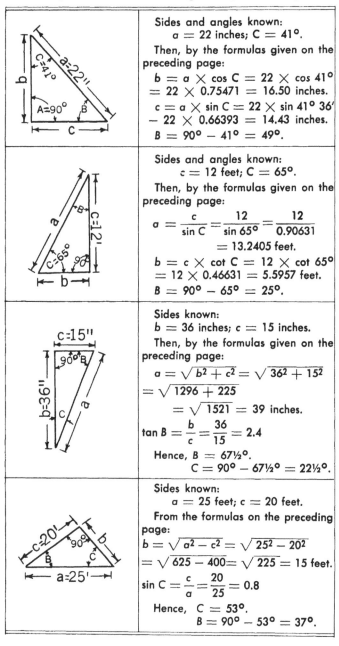

Sides and angles known:
$$a = 22 \text{ inches; } C = 41°.$$
Then, by the formulas given on the preceding page:
$$b = a \times \cos C = 22 \times \cos 41°$$
$$= 22 \times 0.75471 = 16.50 \text{ inches.}$$
$$c = a \times \sin C = 22 \times \sin 41° \, 36'$$
$$- 22 \times 0.66393 = 14.43 \text{ inches.}$$
$$B = 90° - 41° = 49°.$$

Sides and angles known:
$$c = 12 \text{ feet; } C = 65°.$$
Then, by the formulas given on the preceding page:
$$a = \frac{c}{\sin C} = \frac{12}{\sin 65°} = \frac{12}{0.90631}$$
$$= 13.2405 \text{ feet.}$$
$$b = c \times \cot C = 12 \times \cot 65°$$
$$= 12 \times 0.46631 = 5.5957 \text{ feet.}$$
$$B = 90° - 65° = 25°.$$

Sides known:
$$b = 36 \text{ inches; } c = 15 \text{ inches.}$$
Then, by the formulas given on the preceding page:
$$a = \sqrt{b^2 + c^2} = \sqrt{36^2 + 15^2}$$
$$= \sqrt{1296 + 225}$$
$$= \sqrt{1521} = 39 \text{ inches.}$$
$$\tan B = \frac{b}{c} = \frac{36}{15} = 2.4$$
Hence, $B = 67\frac{1}{2}°.$
$$C = 90° - 67\frac{1}{2}° = 22\frac{1}{2}°.$$

Sides known:
$$a = 25 \text{ feet; } c = 20 \text{ feet.}$$
From the formulas on the preceding page:
$$b = \sqrt{a^2 - c^2} = \sqrt{25^2 - 20^2}$$
$$= \sqrt{625 - 400} = \sqrt{225} = 15 \text{ feet.}$$
$$\sin C = \frac{c}{a} = \frac{20}{25} = 0.8$$
Hence, $C = 53°.$
$$B = 90° - 53° = 37°.$$

Solving Problems by Trigonometry

When two straight lines meet, an *angle* is formed between them, and this angle is measured in *degrees*. If the lines intersect perpendicular (at right angles) to each other, as in (a) Fig. 1, the angle is 90 degrees. When the angle is less than 90 degrees, it is called an acute angle, as in (b) Fig. 1.

A *triangle* is a three-sided figure, as in (c) Fig. 1, and when one of the three angles of the triangle is 90 degrees (a right angle), the triangle is called a right angled triangle or a right triangle, as shown in (d) Fig. 1.

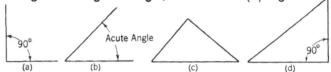

Fig. 1

In right angled triangles, the sum of the three angles always totals 180 degrees; since one angle, by definition, measures 90 degrees, then the sum of the two other (acute) angles equals 180 degrees minus 90 degrees, or 90 degrees.

If the two right triangles are similar in having the same acute angles, then certain relationships or ratios are the same in both triangles, regardless of the lengths of the sides.

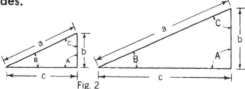

Fig. 2

For example, refer to Fig. 2. In the smaller triangle and the larger triangle, angles A, B, and C are all the same, A being 90°; the triangles differ in the lengths of the respective sides. However, it will be found that these lengths bear the same relation to each other as the similar lengths of the sides of the other triangle. For instance

$$b/a = b_1/a_1$$
and $$c/a = c_1/a_1$$
and $$b/c = b_1/c_1$$

If angle B is 30°, then

        b/a will be .500
        b/c will be .577
        c/a will be .866

and these ratios will hold for both of the triangles in Fig. 2 as well as for any right angled triangle with one acute angle of 30 degrees, regardless of the lengths of the sides.

If b/c is the same (.577) for any 30 degree angle, then it is also true that c/b (called the reciprocal of b/c) will be another figure but also constant; in this case 1 ÷ .577 or 1.73.

All of these side relationships, or ratios, are termed *trigonometric functions* of angle B; and each has been given a name as follows (see Fig. 3):

$$b/a = \text{sine of angle B} = \frac{\text{opposite side}}{\text{hypotenuse}}$$

$$c/a = \text{cosine of angle B} = \frac{\text{adjacent side}}{\text{hypotenuse}}$$

$$b/c = \text{tangent of angle B} = \frac{\text{opposite side}}{\text{adjacent side}}$$

and these are accompanied by the reciprocals

$$a/b = \text{cosecant of angle B} = \frac{\text{hypotenuse}}{\text{opposite side}} = \frac{1}{\text{sine B}}$$

$$a/c = \text{secant of angle B} = \frac{\text{hypotenuse}}{\text{adjacent side}} = \frac{1}{\text{cosine B}}$$

$$c/b = \text{cotangent of angle B} = \frac{\text{adjacent side}}{\text{opposite side}} = \frac{1}{\text{tangent B}}$$

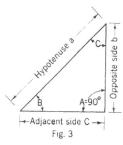

Fig. 3

Now all of these ratios for every degree have been worked out in tables such as that which appears on the pages preceding this explanation. The use of such tables is as follows.

In a right angled triangle, we know that one of the three angles is 90 degrees, and we know that the sum of the other two angles is 90 degrees. Therefore, if we know one of the two acute angles, the other is 90 degrees minus the one we know.

As a result, since we have all the ratios in tables, all we need know about a right angled triangle is the one acute angle and one side and from these the other sides and other angle can be computed. For example, refer to Fig. 4. The one leg of the triangle is 16 inches, and the angle opposite is 27°. What are the other angles and what is the length of a and c?

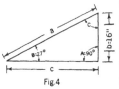

Fig.4

Solution: Since the triangle is right angled, the two acute angles total 90°, so that the angle in the upper right (Angle C) is 90° − 27° or 63°.

The next step is to select any of the ratios previously given which contain b, since that is the side whose length we know. For example, take the first:

$$\text{sine of angle B} = b/a$$

In the table, look up the sine of angle B (which is 27°) and find that the table gives .45399. Now b is given as 16 inches, so that

$$16 \div a = .45399$$

Solving for a, we find that

$$a = 16 \div .45399; \text{ this works out to } 35.2 \text{ inches}$$

Now that we know a, select any ratio which contains b and the other unknown c. For example,

$$\text{tangent of angle B} = b/c$$

and, referring to the table, under tangent and opposite 27°, find .50952, so

$$16 \div c = .50952$$

and solving

$$c = 16 \div .50952 = 31.4 \text{ inches}$$

When any two sides of a right angled triangle are known, it is not necessary to use trigonometry to find the diagonal. The rule is that the sum of the squares of the two sides (or legs) is equal to the square of the diagonal, or hypotenuse.

For example, in the case of Fig. 4, if c = 31.4 inches and b = 16.0 inches what would be the length of a?

First, square 31.4; 31.4 × 31.4 = 985.96. Then square 16 and find this to be 256. Add together and find 985.96 + 256 = 1241.96. This is the square of the hypotenuse a, so extract the square root of 1241.96 and obtain 35.24 inches, the length of a.

To facilitate the solution of squares and square roots, use the tables elsewhere in this book.

The formulas needed to solve right angled triangles with the use of the trigonometric tables are summarized in the two pages preceding this explanation.

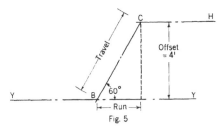

Fig. 5

An example of the use of trigonometry in pipefitting is shown in Fig. 5 where the main YB is to be offset 4 ft. and continue as line CH, with a 60 deg. connection BC connecting the lines. What is the travel, BC?

The angle B is given at 60°, the opposite side is 4 ft.

Use the formula sine B = $\dfrac{\text{opposite side}}{\text{hypotenuse}}$, so that

sine 60° = $\dfrac{4 \text{ ft.}}{\text{travel}}$ The sine of 60° is .866 so that the

travel (or hypotenuse) = $\dfrac{4}{.866}$ = 4.619 ft.

The foregoing can be expressed as

$$travel = \frac{offset}{sine\ 60°} \text{ or}$$

$$travel = offset \times \frac{1}{sine\ 60°}$$

but $\frac{1}{sine\ 60°}$ is equal to the cosecant of 60° so that

travel = offset × cosecant of the angle of bend.

By the same general method, referring to Fig. 5, it is seen that

$$tangent\ 60° = \frac{offset}{run}$$

or $run = offset \times \frac{1}{tan\ 60°}$

but $\frac{1}{tan\ 60°}$ = cotangent 60° so that

run = offset × cotangent of angle of bend.

If the run is known, but not the offset, then

offset = run × tangent of angle of bend

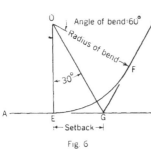

Fig. 6

Setback has been previously defined as the distance measured back from the intersection of the centerlines to the beginning of the bend, as shown in Fig. 6. Note that the setback EG is one side of a right triangle OEG and the angle EOG is one half the angle of bend. Then, since the radius of bend is equal to OE,

$$\frac{setback}{radius\ of\ bend} = \text{tangent of } \tfrac{1}{2} \text{ angle of bend, or}$$

setback = radius of bend × tangent ½ the angle of bend.

Angles on the Steel Square

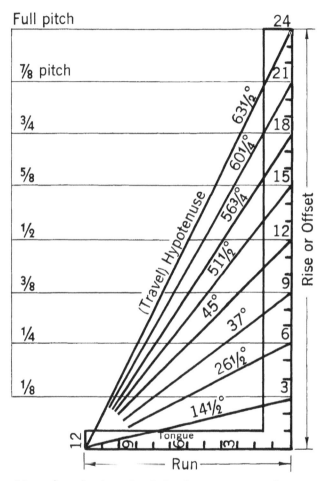

Note that the length of the hypotenuse is the same as the dimension "travel" in piping offset calculations. Pitch is the fraction of the full blade used in setting the angle from the tip of the tongue.

The blade is the longer, wider leg of the steel square. Measurements for constructing angles are taken on the outside edge of the square.

Angles and Length of Hypotenuse
on the Steel Square

When the Run is 12 Inches (1 Ft.)		
And the Rise in Inches Is:	the Angle in Degrees Is:	Length of the Hypotenuse is the Run Times:
1¹⁄₁₆	5	1.0038
1⁹⁄₁₆	7½	1.0086
2⅛	10	1.0154
2¹¹⁄₁₆	12½	1.0243
3³⁄₁₆	15	1.0353
3¹¹⁄₁₆	17½	1.0457
4⅜	20	1.0642
5	22½	1.0824
5⅝	25	1.1034
6¼	27½	1.1274
6¹⁵⁄₁₆	30	1.1547
7⅝	32½	1.1857
8⅜	35	1.2208
9¹⁄₁₆	37½	1.2521
10¹⁄₁₆	40	1.3054
11	42½	1.3563
12	45	1.4142
13⅛	47½	1.4802
14⁵⁄₁₆	50	1.5557
15⅝	52½	1.6427
17⅛	55	1.7434
18¹³⁄₁₆	57½	1.8611
20¹³⁄₁₆	60	2.0000
23¹⁄₁₆	62½	2.1657

Laying Out Angles by Use of Six-Foot Rule

Reasonably accurate angles can be laid out by using a six-foot folding rule as illustrated in the sketch. When the tip B is held at the dimension indicated in the table below, the angle formed at X is as shown. For example, when the tip is at $22\frac{1}{4}$ inches, the angle X is 60 degrees.

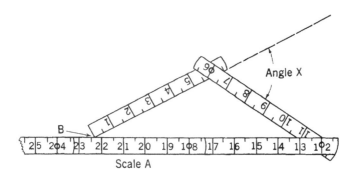

Scale A

Angle X, Deg.	Set Tip B at	Angle X, Deg.	Set Tip B at
15	$23\frac{7}{8}$	50	$22\frac{11}{16}$
$17\frac{1}{2}$	$23\frac{13}{16}$	60	$22\frac{1}{4}$
20	$23\frac{3}{4}$	65	$21\frac{15}{16}$
$22\frac{1}{2}$	$23\frac{11}{16}$	70	$21\frac{5}{8}$
25	$23\frac{5}{8}$	75	$21\frac{5}{16}$
$27\frac{1}{2}$	$23\frac{9}{16}$	90	$20\frac{1}{4}$
30	$23\frac{1}{2}$	$112\frac{1}{2}$	$18\frac{3}{8}$
35	$23\frac{3}{8}$	120	$17\frac{5}{8}$
40	$23\frac{3}{16}$	135	$16\frac{3}{16}$
45	$22\frac{15}{16}$	150	$14\frac{11}{16}$

Pipefitters Dictionary

ALLOY PIPE. A steel pipe with one or more elements other than carbon which give it greater resistance to corrosion and more strength than carbon steel pipe.

ANGLE OF BEND. In a pipe, the angle at the center of the bend between radial lines from the beginning and end of the bend to the center.

ANGLE VALVE. A valve, usually of the globe type, in which the inlet and outlet are at right angles.

BACKING RING. A metal strip used to prevent melted metal from the welding process from entering a pipe when making a butt-welded joint.

BELL AND SPIGOT JOINT. The commonly used joint in cast-iron pipe. Each piece is made with an enlarged diameter or bell at one end into which the plain or spigot end of another piece is inserted when laying. The joint is then made tight by cement, oakum, lead, or rubber calked into the bell around the spigot.

BLACK PIPE. Steel pipe that has not been galvanized.

BLANK FLANGE. A flange in which the bolt holes have not been drilled.

BLIND FLANGE. A flange used to seal off the end of a pipe.

BONNET. Part of a valve used to guide and support the valve stem.

BRANCH. The outlet or inlet of a fitting not in line with the run, and taking off at an angle with the run.

BRANCH TEE. A tee having many side branches.

BRAZED. Joined by hard solder.

BULL HEAD TEE. A tee the branch of which is larger than the run.

BUSHING. A pipe fitting for connecting a pipe with a female fitting of larger size. It is a hollow plug with internal and external threads.

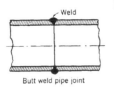

Butt weld pipe joint

BUTT WELD JOINT. A welded pipe joint made with the ends of the two pipes butting each other, the weld being around the periphery.

BUTT WELD PIPE. Pipe welded along a seam butted edge to edge and not scarfed or lapped.

BYPASS.   In a pipe line, a supplementary line leaving the main run and rejoining it at some point beyond a valve or other apparatus so that service is not interrupted when the valve or apparatus is not usable.

CARBON STEEL PIPE.   Steel pipe which owes its properties chiefly to the carbon which it contains.

CHECK VALVE. A valve designed to allow a fluid to pass through in one direction only. A common type has a plate so suspended that the reverse flow aids gravity in forcing the plate against a seat, shutting off reverse flow.

CLOSE NIPPLE.   A nipple with a length twice the length of a standard pipe thread.

COMPANION FLANGE.   A pipe flange to connect with another flange or with a flanged valve or fitting. It is attached to the pipe by threads, welding, or other method and differs from a flange which is an integral part of a pipe or fitting.

COMPRESSION JOINT.   A multi-piece joint with cup shaped threaded nuts which, when tightened, compress tapered sleeves so that they form a tight joint on the periphery of the tubing they connect.

CONDENSATE.   When steam loses sufficient heat it returns to water; this liquefied steam is condensate.

COUPLING.   A threaded sleeve used to connect two pipes. They have internal threads at both ends to fit external threads on pipe.

CROSS.   A pipe fitting with four branches in pairs, each pair on one axis, and the axes at right angles.

CROSS-OVER.   A small fitting with a double offset, or shaped like the letter U with the ends turned out. It is only made in small sizes and used to pass the flow of one pipe past another when the pipes are in the same plane.

CROSS VALVE.   A valve fitted on a transverse pipe so as to open communication between two parallel pipes.

Cup weld

CUP WELD. A pipe weld where one pipe is expanded on the end to allow the entrance of the end of the other pipe. The weld is then circumferential at the end of the expanded pipe.

DOUBLE EXTRA-STRONG PIPE. A schedule of steel or wrought iron pipe weights in common use.

DOUBLE SWEEP TEE. A tee made with easy (long radius) curves between body and branch.

DROP ELBOW. A small ell used in gas fitting. These fittings have wings cast on each side, the wings having countersunk holes so that they may be fastened by wood screws to a ceiling, wall, or framing timbers.

DROP TEE. One having the wings of the same type as the drop elbow.

ELBOW (Ell). A fitting that makes an angle between adjacent pipes. The angle is 90 degrees, unless another angle is specified.

EXPANSION JOINT. A joint whose primary purpose is not to join pipe but to absorb that longitudinal expansion in the pipe line due to heat.

EXPANSION LOOP. A large radius bend in a pipe line to absorb longitudinal expansion in the line due to heat.

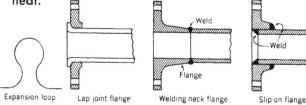

Expansion loop   Lap joint flange   Welding neck flange   Slip-on flange

FLANGE. In pipe work, a ring-shaped plate on the end of a pipe at right angles to the end of the pipe and provided with holes for bolts to allow fastening the pipe to a similarly equipped adjoining pipe. The resulting joint is a flanged joint.

FLANGE FACES. Pipe flanges which have the entire face of the flange faced straight across, and use either a full face or ring gasket, are commonly employed for pressures less than 125 pounds on steam and water lines.

FUSION WELD.   Joining metals by fusion, using oxy-acetylene or electric arc.

GALVANIZED PIPE.   Steel pipe coated with zinc to resist corrosion.

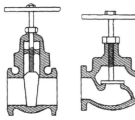

Gate valve   Globe valve

GATE VALVE.   A valve employing a gate, often wedge-shaped, allowing fluid to flow when tne ၄...e is lifted from the seat. Such valves have less resistance to flow than globe valves.

GLOBE VALVE.   One with a somewhat globe shaped body with a manually raised or lowered disc which when closed rests on a seat so as to prevent passage of a fluid.

GROUND JOINT.   One where the parts to be joined are precisely finished and then ground in so that the seal is tight.

HEADER.   A large pipe or drum into which each of a group of boilers is connected. Also used for a large pipe from which a number of smaller ones are connected in line and from the side of the large pipe.

LAPPED JOINT.   A pipe joint made by using loose flanges on lengths of pipe whose ends are turned over or lapped over to produce a bearing surface for a gasket or metal-to-metal joint.

LAP WELD PIPE.   That made by welding along a scarfed longitudinal seam in which one part is over-lapped by the other.

LEAD JOINT.   A joint made by pouring molten lead into the space betwen a bell and spigot, and making the lead tight by calking.

LIP UNION.   A form of union characterized by the lip that prevents the gasket from being squeezed into the pipe so as to obstruᴄr the flow.

MALLEABLE IRON.   Cast iron heat-treated to reduce its brittleness. The process enables the material to stretch to some extent and to stand greater shock.

MANIFOLD.   A fitting with a number of branches in line connecting to smaller pipes. Used largely as an interchangeable term with header.

MEDIUM PRESSURE.   When applied to valves and fittings, implies they are suitable for a working pressure of from 125 to 175 pounds per square inch.

MILL LENGTH.   Also known as random length. Run-of-mill pipe is 16 to 20 ft. in length. Some pipe is made in double lengths of 30 to 35 ft.

NEEDLE VALVE.   A valve provided with a long tapering point in place of the ordinary valve disk. The tapering point permits fine graduation of the opening.

NIPPLE.   A tubular pipe fitting usually threaded on both ends and under 12 inches in length. Pipe over 12 inches long is regarded as cut pipe.

O. D. PIPE.   Smaller size pipe is usually designated by its inside diameter. Over 14 inches, however, the nominal size is the outside diameter, and such pipe is termed O. D.

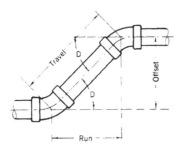

OFFSET.   When, to avoid an obstruction or some other reason, a line of pipe is to be continued as a line running in the same direction, the perpendicular distance between the centerline of the one and the extension of the other is the offset. Travel is the length of the connecting line measured between centerline intersections, and run is the other leg of the triangle formed by the travel and the offset.

PLUG VALVE.   One with a short section of a cone or tapered plug through which a hole is cut so that fluid can flow through when the hole lines up with the inlet and outlet, but when the plug is rotated 90°, flow is blocked.

REDUCER.   A fitting with a larger size at one end than at the other; the larger size is designated first. Reducers are threaded inside, unless specified flanged, welded, or for some special joint.

RELIEF VALVE. One designed to open automatically to relieve excess pressure.

RESISTANCE WELD PIPE. Pipe made by bending a plate into circular form and passing electric current through the material to obtain a welding heat.

ROLLING OFFSET. Same as offset, but used where the two lines are not in the same vertical or horizontal plane.

ROTARY PRESSURE JOINT. A joint for connecting a pipe under pressure to a rotating machine.

RUN. A length of pipe made of more than one piece of pipe; a portion of a fitting having its ends in line or nearly so, in contradistinction to the branch or side opening, as of a tee. In this book, the term run has still another meaning, as explained under Offset.

SADDLE FLANGE. A flange curved to fit a boiler or tank and to be attached to a threaded pipe. The flange is riveted or welded to the boiler or tank.

SATURATED STEAM. Steam at the same temperature as water boils under the same pressure.

SCREWED FLANGE. A flange screwed on the pipe which it is connecting to adjoining pipe.

SCREWED JOINT. A pipe joint consisting of threaded male and female parts screwed together.

SEAMLESS PIPE. Pipe or tube formed by piercing a billet of steel and then rolling.

SERVICE FITTING. A street ell or street tee with male threads at one end and female threads at the other.

SERVICE PIPE. A pipe connecting water or gas mains with a building.

SET. Same as offset, but also used in place of offset where the connected pipes are not in the same vertical or horizontal plane—a rolling offset.

SETBACK. In a pipe bend, the distance measured back from the intersection of the centerlines to the beginning of the bend.

SHORT NIPPLE. One whose length is a little greater than that of two threaded lengths or somewhat longer than a close nipple so that it has some unthreaded portion between the two threads.

SHOULDER NIPPLE.   A nipple of any length which has a portion of pipe between two pipe threads. As generally used, however, it is a nipple halfway between the length of a close nipple and a short nipple.

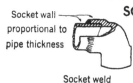

Sleeve weld

SLEEVE WELD.   A joint made by butting two pipes together and welding a sleeve over the outside.

SLIP-ON FLANGE.   A flange slipped over the end of the pipe and then welded to the pipe.

Socket wall —
proportional to
pipe thickness

Socket weld

SOCKET WELD.   A joint made by use of a socket weld fitting which has a prepared female end or socket for insertion of the pipe to which it is welded.

SOLDER JOINT.   A method of joining tube by use of solder.

SPIRAL PIPE.   Pipe made by coiling a plate into a helix and riveting or welding the overlapped edges.

STAINLESS STEEL PIPE.   An alloy steel pipe with corrosion-resisting properties, usually imparted by nickel and chromium.

STANDARD PRESSURE.   Formerly used to designate cast-iron flanges, fittings, valves, etc., suitable for a maximum working steam pressure of 125 lb.

STREET ELBOW.   An elbow with male thread on one end, and female thread on the other end.

SUPERHEATED STEAM.   Steam at a higher temperature than that at which water would boil under the same pressure.

SWING JOINT.   An arrangement of screwed fittings and pipe to provide for expansion in pipe lines.

SWIVEL JOINT.   One employing a special fitting designed to be pressure tight under continuous or intermittent movement of the machine or part to which it is connected.

TEE.   A fitting, either cast or wrought, that has one side outlet at right angles to the run.

TRAVEL.   See Offset.

UNION. A device used to connect pipes and usually consisting of three pieces: the thread end fitted with exterior and interior threads; second, the bottom end fitted with interior threads and a small exterior shoulder; and third, the ring which has an inside flange at one end while the other end has an inside thread like that on the exterior of the thread end. Unions are extensively used because they permit of connections with little disturbance of the pipe positions.

UNION ELL. An ell with a male or female union at one end.

UNION JOINT. A pipe coupling, usually threaded, which permits disconnection without disturbing other sections.

UNION TEE. A tee with male or female union at one end of the run.

WELDING-END VALVES. Valves without end flanges and with ends tapered and beveled for butt welding.

WELDING FITTINGS. Wrought or forged-steel prefabricated elbows, tees, reducers, saddles, and the like, beveled for welding to pipe.

WELDING NECK FLANGE. A flange with a relatively long neck beveled for butt welding to the pipe.

WIPED JOINT. A lead pipe joint in which molten solder is poured upon the desired place, after scraping and fitting the parts together, and the joint is wiped up by hand with a moleskin or cloth pad while the metal is in a plastic condition.

WROUGHT IRON. Iron refined to a plastic state in a puddling furnace. It is characterized by the presence of about 3 per cent of slag irregularly mixed with pure iron and about 0.5 per cent carbon.

WROUGHT PIPE. This term refers to both wrought steel and wrought iron. Wrought in this sense means worked, as in the process of forming furnace-welded pipe from skelp, or seamless pipe from plates or billets. The expression wrought pipe is thus used as a distinction from cast pipe. When wrought-iron pipe is referred to, it should be designated by its complete name.

WYE (Y). A fitting, either cast or wrought, that has one side outlet at any angle other than 90 degrees.

INDEX

ABS plastic pipe, 318

Adapters, cast brass, 285, 286

Alloys, composition of brazing, 290

Alloys, melting temperatures of, 288

Angles:
  for bending, 1
  for mitering, 101
  on the six-foot rule, 395
  on the steel square, 393
  trigonometric functions of, 381-385

Angle valves:
  definition of, 305
  flanged, 250 to 254
  welding, 189

Apothecaries' measure, 366

Apothecaries' weight, 367

Arc, length of in bend, 1

Areas and circumferences of circles, 339-344

Areas and volumes of various shapes, 359-363

Avoirdupois weight, 367

Bending pipe around a circle, 85

Bending pipe, general 1 to 89

Bending plastic pipe, 310

Bending radius, minimum, 88

Bending 90° turns, 81 to 84

Bends:
  circular, 78
  crossover, 75
  expansion, 79
  fabrication of, 80
  offset, 54 to 71
  offset U, 76, 77
  setback and length of, 2 to 53
  wrinkle, 89

Brackets, mitered, 115 to 120

Brazing, 287

Brazing alloys, 290

Butt welding fittings, dimensions of, 173 to 192

Cast brass:
  adapters, 285, 286

elbows, 283, 284

fittings, laying lengths of, 270 to 286

tees, 283, 284

Cast iron:
  check valves, flanged, 255 to 259
  fittings, dimensions of, 141 to 149
  fittings, laying lengths of, 163, 167
  flanged fittings, 200 to 212
  valves, flanged, 244, 245, 250, 256

Cellulose acetate butyrate pipe, 320

Centigrade-Fahrenheit conversion, 369

Check valves:
  definition of, 306
  flanged, 255 to 259
  trouble shooting for, 308
  welding, 191

Chemicals, effect of on metal pipe, 328 to 331

Chemical resistance of plastic pipe, 328

Circle (s):
  bending pipe around, 85
  circumferences and areas of, 339
  dividing, into segments, 192
  mitering pipe around, 114
  screwed pipe around, 100

Circular bends, 78

Circumference divided into chords, 193

Circumference of pipe, dividing, 121
  tables of, 130 to 136

Circumferences and areas of circles, 339

Clearance for turning screwed fittings, 158

Coils:
  bending, 85
  screwed, 100
  welded, 114

Color identification of piping, 301

Contents of cylindrical tanks, 297

Conversion(s):
    cubic feet to gallons, 376
    factors, various, 378
    Fahrenheit-centigrade, 369
    feet of water to pounds per
        square inch, 374
    gallons to cubic feet, 377
    inches to decimals of a foot,
        368
    psi to feet of water, 375
    tables, 368 to 379
Copper tube:
    expansion of, 296
    dimensions of:
        Type DWV, 137
        Type K, 134
        Type L, 135
        Type M, 136
Corrosion resistance of plastic
    and metal pipes, 324, 328
Cosecants, table of, 381-385
Cosines, table of, 381-385
Cotangents, table of, 381-385
Couplings:
    cast iron reducer, laying
        length of, 166
    reducing, solder joint, 282
    screwed reducing, dimen-
        sions of, 155
    solder joint, 274
Crosses:
    cast iron screwed,
        dimensions of, 141, 145
    laying lengths of, 163
    flanged, 201 to 243
    welding, 184 to 186
Crossover bends, 72, 75
Cubic measure, 366

Decimal equivalents of frac-
    tions, 123
Decimals of a foot, in inches,
    368
Dictionary, pipefitters,
    396-403
Dies, threading, 138
Dimensions of:
    ABS pipe, 319
    butyrate pipe, 321
    butt welding fittings,
        173 to 192
    cast iron fitings, 141 to 149
    flanged fittings, 193 to 269
    flat gaskets, 266, 267
    pipe nipples, 157

polyethylene pipe, 316
PVC pipe, 313
ring joint gaskets,
    262 to 265
Schedule 40 steel pipe, 130
Schedule 120 steel pipe,
    132
solder joint fittings,
    270 to 286
stainless steel pipe, 133
Type DWV copper tube,
    137
Type K copper tube, 134
Type L copper tube, 135
Type M copper tube, 136
U bolts for pipe hangers,
    291
Dividing circumference of pipe,
    121
Drainage tube, copper, 137
Dry measure, 366

Elbows:
    cast brass, one end
        threaded, 283, 284
    cast brass, solder joint,
        271, 272
    cast iron, laying lengths of,
        163, 165
    cast iron screwed, dimen-
        sions of, 141 to 144
    flanged, 200 to 243
    screwed reducing,
        143 to 145
    welding, dimensions of, 173
    welding, reducing, 178, 179
Ends, cast brass, 270
    butt welding, 182
Engagement, pipe thread, 140
Equivalents, decimal,
    of fractions, 123
Expansion bends, 79
Expansion of metal pipe, 296
Expansion of plastic pipe, 312

Fahrenheit-centigrade
    conversion, 369
Fittings:
    flanged, 193 to 269
    screwed, 141 to 172
    solder joint, 270 to 286
    turning clearance for,
        158 to 162
    welding, 173 to 192

Flanged fittings, laying lengths of, 200 to 261

Flanges, ring joint, laying length of, 260

Flanges, templates for drilling, 193 to 199

Fluxes for soldering, 289, 290

Formulas, useful, 380

Fractions, decimal equivalents of, 123

Gallons to cubic feet, 377

Gallons of water in pipe, 295

Gaskets:
flat, 266, 267
ring joint, 262 to 265
types of, 268

Gate valves:
definition of, 304
flanged, 244 to 249
trouble shooting for, 307
welding, 187

Globe valves:
definition of, 305
flanged, 250 to 254
trouble shooting for, 308
welding, 189

Hangers, pipe, U bolts for, 291
spacing of, 292

Identification of piping systems, 300

Joining plastic pipe, 310

Joints, mitered, 101 to 129
soldering, 287
welded, 101 to 129

Laying lengths of:
butt welding fittings, 173 to 192
flanged fittings, 200 to 261
flanged valves, 244 to 259
screwed fittings, 163 to 172
solder joint elbows and tees, with pipe thread ends, 283, 284
solder joint fittings, 270 to 286

Length of chords, circle, 193

Length of pipe in bend, 1
see also Setback and length of bend

Liquid measure, 366

Malleable iron fittings:
dimensions of, 141, 142, 150 to 156
laying lengths of, 163 to 172

Measure:
apothecaries', 366
cubic, 366
dry, 366
length, 365
liquid, 366
nautical, 365
pressure, 367
shipping, 366
square, 365
surveyor's, 365
wire, 365

Melting points and weights of metals, 337

Melting temperatures of soldering alloys, 288

Metal sheets, gages of, 336

Metals:
soldering of, 290
spark tests for, 334
weights and melting points of, 337

Minimum bending radius, 88

Mitered brackets, 115 to 120

Mitered pipe joints, 101

Miter templates, construction of, 124

Nautical measure, 365

Nipples, pipe, dimensions of, 157

Numbers, reciprocals of, 364

Numbers, squares and square roots of, 345-358

Offset bends:
15°, 56
15°, two pipes, 57
22½°, 58
22½°, two pipes, 59
30°, 60
30°, two pipes, 61
45°, 62
45°, two pipes, 63
60°, 64
60°, two pipes, 65
72°, 66
72°, two pipes, 67
90°, one or more pipes, 68

Offset bends, continuous, 69
Offset bends, double, 70
Offset bends, run in, 54
Offset bends, travel in, 54
Offset quarter bends, 210°, 74
Offsets, screwed, travel and run center to center of:
  11¼° elbows, 92
  22½° elbows, 93
  30° elbows, 94
  45° elbows, 95
  60° elbows, 96
  72° elbows, 97
Offsets, rolling, 98
Offsets, screwed, 90
Offsets, screwed, two or more pipes, 91
Offset U bend, 76, 77

Pipe:
  ABS plastic, 318
  bending of metal, 1
  bending of plastic, 310
  CAB plastic, 320
  expansion of metal, 296
  expansion of plastic, 312
  length of in bend, 1
  metal, chemical resistance of, 329
  minimum bending radius for, 88
  mitering of, 101
  plastic, 309 to 333:
    chemical resistance of, 328
    corrosion resistance of, 324, 328
    support spacing for, 322, 323
  polyethylene, 315
  Schedule 40 steel, dimensions of, 130
  Schedule 120 steel, dimensions of, 132
  stainless steel, dimensions of, 133
  steel, threads for, 139
  thread engagement of, 140
  threading of, 138
  volume of water in, 295
  weight of water in, 294
Pipe bending, metal, 1
Pipe bending, plastic, 310
Pipe circumference, dividing, 121

Pipe coils:
  bending, 85
  screwed, 100
  welded, 114
Pipe expansion, metal, 296
Pipe expansion, plastic, 312
Pipe hangers, U bolts for, 291
  spacing of, for metal pipe, 292
  spacing of, for plastic pipe, 311
Pipe nipples, dimensions of, 157
Pipefitters dictionary, 396-403
Pipe plastics, characteristics of, 325
Pipe supports, spacing of, 292, 311
Pipe thread engagement, 140
Piping systems, identification of, 300
Plastic pipe, 309 to 333:
  corrosion resistance of, 324, 328
  dimensions of, 313 to 321
  properties of, 325 to 327, 332
  support spacing for, 322, 323
Polyethylene pipe, 315
Polyvinyl chloride pipe, 309
  dimensions and pressure limits of, 313
Pressure:
  conversion table for, 374-375
  measures of, 367
  and temperature of steam, 338
Pressure limits of PVC pipe, 313
Pressure limits of polyethylene pipe, 316

Radius multipliers for bending, 1, 46
Reciprocals of numbers, 364
Reducer couplings, 155, 166, 170
Reducer couplings, solder joint, 274
Reducers, welding, 180
Reducing elbows:
  cast brass, solder joint, 273
  cast iron, screwed, 143
  flanged, 200
  welding, 178, 179

Reducing tees:
    cast iron, screwed, 147, 156
    solder joint, 275 to 281
    welding, 175
Returns, welding, 181
Right angle triangles, 386
Ring joint flanges, 260
Ring joint gaskets, 262
Rolling offsets, 98
Run in offset bends, 54

Screwed fittings:
    clearance for turning, 158
        to 162
    dimensions of, 141 to 157
    laying lengths of,
        163 to 172
Screwed offsets, 90
Screwed turns, 99
Secants, table of, 381-385
Segments of pipe circum-
    ference, 121, 122
Setback, 2
Setback and length of:
    15° bends, 3 to 6
    22½° bends, 7 to 10
    30° bends, 11 to 14
    45° bends, 15 to 18
    60° bends, 19 to 22
    72° bends, 23 to 26
    90° bends, 27 to 30
    112½° bends, 31 to 33
    120° bends, 34 to 36
    135° bends, 37 to 39
    150° bends, 40 to 42
    180° bends, 43, 44
    bends, any radius, ¼° to
        180°, 45 to 53
Sheet metal gages, 336
Shipping measure, 366
Sines, table of, 381 to 385
Solder joint adapters, 285, 286
Solder joint fittings, laying
    lengths of, 270 to 286
Solder, melting temperatures
    of, 288
Solders for various metals,
    290
Soldering and brazing, 287
Spacing flange bolt holes, 193
Spacing of pipe supports, 292
Spark tests for metals, 334
Square, angles on steel, 393
Square measure, 365
Squares and square roots of
    numbers, 345-358

Standard pipe, minimum
    bending radius for, 88
    dimensions of, 130
Steam, saturated properties,
    338
Steel flanged fittings,
    213 to 243
Steel pipe, expansion of, 296
Steel pipe, stainless, dimen-
    sions of, 133
Steel pipe, threads for, 139
Steel square, angles of, 393
Steel and wrought iron pipe,
    dimensions of, 130
Street elbows, 169, 172
Street elbows, cast brass,
    solder joint, 273
Stub end, welding, 182
Supports, spacing of metal
    pipe, 292
Supports, spacing of plastic
    pipe, 322, 323
Surveyor's measure, 365
Sweating, 290
Systems, piping, identification
    of, 300

Tangents, table of, 381-385
Tanks:
    bending pipe around, 85
    contents of cylindrical, 297
    screwed coils around, 100
    welding pipe around, 114
Tap drill sizes, 139
Tees:
    cast brass, 283, 284
    cast brass, solder joint, 271
    cast iron, dimensions of,
        141, 142
    flanged, 201 to 243
    laying lengths of screwed,
        163
    reducing outlet, screwed,
        147 to 149, 156
    reducing, solder joint,
        275 to 281
    welding, dimensions of,
        174 to 177
Temperature of saturated
    steam, 338
Temperatures, melting,
    of alloys, 288
Templates, 45° branch, 126
Templates for drilling flanges,
    193 to 199

Templates, miter, construction of, 124

Templates, tee branch, 125

Templates, for true wye, 127

Tests, spark, for metals, 334

Thermoplastics, physical properties of, 332

Thermoplastic pipe, 309

Thickness of steel sheets, 336

Thread engagement, pipe, 140

Threading pipe, 138

Threading plastic pipe, 310

Threads for steel pipe, 139

Triangles, right, formulas for, 386

Triangles, trigonometry of, 388-392

Trigonometric functions, 381-385

Trigonometry, solving problems by, 388-392

Travel in offset bends, 54

Travel and run in screwed offsets, 90

Trouble shooting for valves, 307, 308

Tube, copper:
dimensions of, 134 to 137
expansion of, 296
volume of water in, 295
weight of water in, 294

Turning clearance for screwed fittings, 158 to 162

Turns, 90° bends, 81

Turns, welded, 102 to 113

Types of gaskets, 268

U bolts for pipe hangers, 291

Valves:
angle, definition of, 305
check, definition of, 306
gate, definition of, 304
gate, flanged, 244 to 249
gate, welding, laying lengths of, 187, 188
globe and angle, flanged, 250 to 254
globe and angle, welding, laying lengths of, 189
globe, definition of, 305
swing check, flanged, 255 to 259
swing check, welding, 191
trouble shooting for, 307, 308

Volumes and areas, 359-363

Volume of tanks, 297

Volume of water in pipe, 295

Water, gallons of, in pipe, 295
weight of, in pipe, 294

Weight(s):
apothecaries', 367
avoirdupois, 367
and measures, 365-367
and melting points of metals, 337
of steel sheets, 336
of water in pipe, 294

Welded branch connections:
field marking, 129
templates for, 125 to 128

Welded pipe joints, 101

Welding fittings, 173 to 192

Welding plastic pipe, 310

Wire measure, 365

Wrinkle bends, 89

Wrought iron, expansion of, 296

Wrought iron pipe, dimensions of, 130

Wyes, cast iron, screwed, laying lengths of, 164
templates for, 127

Lightning Source UK Ltd.
Milton Keynes UK
UKHW010618201119
353895UK00001B/191/P